# Mapping and Scheduling Algorithms for Synchronized Individual Production

# Zuordnungs- und Terminierungsalgorithmen für die synchronisierte Einzelfertigung

Von der Fakultät für Maschinenwesen
der Rheinisch-Westfälischen Technischen Hochschule Aachen
zur Erlangung des akademischen Grades eines
Doktors der Ingenieurwissenschaften
genehmigte Dissertation

vorgelegt von

Sean Edward Humphrey

**Berichter:**

Univ.-Prof. Dr.-Ing. Robert Heinrich Schmitt
Außerplanmäßiger Professor Dr.-Ing. Thomas Prefi

Tag der mündlichen Prüfung: 12. Juli 2016

**Bibliografische Information der Deutschen Nationalbibliothek**

Die Deutsche Nationalbibliothek verzeichnet diese Publikation in der Deutschen Nationalbibliografie; detaillierte bibliografische Daten sind im Internet über http://dnb.ddb.de abrufbar.

Sean Edward Humphrey

Mapping and Scheduling Algorithms for Synchronized Individual Production

1. Auflage 2016

Umschlagseite gestaltet von Dr. Alexander van Laack

© 2016 van Laack GmbH, Aachen, Buchverlag

Roermonder Str. 312, 52072 Aachen

Internet: www.van-Laack.de, E-Mail: info@van-Laack.de

Druck und Vertrieb durch:

Books on Demand (BoD) GmbH, In de Tarpen 42, 22848 Norderstedt, www.bod.de

Printed in Germany

ISBN 978-3-936624-35-9

D 82 (Diss. RWTH Aachen University, 2016)

Dedicated to my parents and grandparents

# Preface

The research underlying this thesis took place primarily during my time as a research associate at the Laboratory for Machine Tools and Production Engineering (WZL) of the RWTH Aachen University. I am profoundly grateful to Professor Robert Schmitt, who gave me the opportunity to pursue my doctoral studies in an environment both rich with challenging issues from industrial practice and conducive to their solution in the context of academic research.

I would also like to thank Professor Thomas Prefi for agreeing to serve as co-examiner of my thesis. My thanks also go to Professors Hopmann and Schomburg for chairing the defense proceedings and serving as an observer respectively.

Over the years many friends and colleagues showed interest in the topic of my dissertation and the challenges I encountered. I hereby would like to express my appreciation for engaging me in countless stimulating discussions, challenging my perceptions and providing helpful fresh perspectives. Specifically, I would like to express my gratitude to Dr. Sebastian Schmitt for his advice and continuous feedback – especially in the area of research methodology – from the first idea to the finished thesis. Moreover, I would like to thank Simon Müller and Max Ellerich for innumerable valuable discussions on encountered mathematical challenges and for their unwavering commitment when implementing the synchroTecS software. My thanks go to Dr. Alexander van Laack, for his help in finalizing the dissertation and assistance in preparing for its defense. I would also like to recognize my predecessor Hagen Ziskoven for laying an excellent foundation in the field of synchronized individual production and for his insightful input in the various stages of my research. I would like to acknowledge my countless unnamed colleagues and student research assistants at the WZL for their support, for creating a unique fun working environment and making the last five years a time I will always look back upon fondly.

I would also like to convey my appreciation for the insightful, collaborative discussions with industrial experts. These were instrumental to collecting requirements from industrial practice, designing the planning algorithm and validating the results on the factory floor. Without them I could not have appreciated the complexity and challenges resulting from a practical application. In this context I would explicitly like to mention: Herbert Johann, Gerd

Ringelmann, Ulrich Rossmehl, Dominik Schüler, Dr. Christian Hinsel, Andreas Nieberle, Renée Langlotz, Jochem Rotthaus and Jürgen Kleinfeld from ZF Friedrichshafen AG, Hirschvogel Umformtechnik and SONA AutoComp Germany GmbH.

Last but not least I would like to thank my girlfriend Antonia Fels, whose patience, critical feedback and continuous encouragement during the conception of this dissertation was invaluable. Finally, my deepest gratitude goes to my parents, who not only continuously reviewed this thesis as it was being written but also offered essential guidance in my formative years, actively supported my education and thus gave me the best start in life I could have hoped for. It is to them I dedicate this dissertation.

Aachen, August 2016                                                    Sean Humphrey

# Zusammenfassung

Der Werkzeugbau versorgt produzierende Unternehmen mit Werkzeugen und Sondermaschinen für die Serienproduktion. Somit besetzt er eine Schlüsselposition in der Wertschöpfungskette. Angesichts zunehmenden Kostendrucks seitens Mitbewerbern aus Niedriglohnländern ist technische Produktqualität jedoch nicht mehr ausreichend, um langfristig wirtschaftlich erfolgreich agieren zu können. Vor dem Hintergrund gesteigerter Variantenvielfalt und verkürzter Produktlebenszyklen werden die Aspekte Effizienz, kurze Durchlaufzeiten und Termintreue zunehmend wichtiger. Um diesen Markforderungen zu entsprechen, wurde das Synchronisationsprinzip auf die Fertigung dieser hochindividualisierten Produkte, die bislang vielfach nach dem Werkstattprinzip hergestellt werden, angewandt. In der Praxis konnten durch Anwendung dieses Produktionsprinzips erhebliche wirtschaftliche Potentiale gehoben, Abläufe auf dem betrieblichen Hallenboden vereinfacht und ein transparenter Überblick über das Produktionsgeschehen geschaffen werden. Diesen Vorteilen standen jedoch hohe Komplexität und Aufwände in den notwendigen Planungsaktivitäten gegenüber.

In dieser Dissertation werden Algorithmen zur Automatisierung der zwei primären Planungsaktivitäten für die synchronisierte Einzelfertigung vorgestellt: die Zuordnung von Aufträgen zu einer Sequenz synchronisierter Produktionslinien und die zeitliche Ablaufplanung der zugeordneten Aufträge. Beide Algorithmen berücksichtigen hierbei die im Konflikt stehenden Ziele Termintreue, minimale Durchlaufzeiten und maximale Maschinenauslastung.

Zur Sicherstellung der industriellen Anwendbarkeit wurden die Forschungsaktivitäten in Zusammenarbeit mit drei großen deutschen Automobilzulieferern durchgeführt. Nach der Anforderungsaufnahme in Experteninterviews wurden die Algorithmen in der Entwurfsphase theoretisch geprüft und abschließend mittels einer Softwareimplementierung prototypisch auf dem betrieblichen Hallenboden validiert. Nach Abschluss der in dieser Arbeit beschriebenen Forschungsarbeiten wurde die entwickelte Softwarelösung zur Serienreife weiterentwickelt und befindet sich nun als primäres Feinplanungssystem bei einigen Unternehmen bereits im täglichen Einsatz.

# Outline

# Table of Contents

# Abbreviations & Variables

$TSC$    Tactstation Container. 67

$TSM_{AMI}$    Absolute maximum of Isolated Manufacturing Steps allowed in a Tactsequence generated during Tactsequence Mapping. $TSM_{AMI} \in \mathbb{N}$. 71

$TSM_{API}$    Allow generation of Tactsequences consisting purely of Isolated Manufacturing Steps. $TSM_{API} \in \{0, 1\}$ with 0 to disable and 1 to enable. 71

$TSM_{ATSD}$    Maximum absolute Tactsequence dilation allowed in a Tactsequence generated during Tactsequence Mapping. $TSM_{ATSD} \in \mathbb{N}$. 72

$TSM_{MLO}$    Maximum number of logistics operations allowed in a Tactsequence generated during Tactsequence Mapping. $TSM_{MLO} \in \mathbb{N}$. 73

$TSM_{RTSD}$    Maximum relative Tactsequence dilation allowed in a Tactsequence generated during Tactsequence Mapping. $TSM_{RTSD} \in \mathbb{R}, TSM_{RTSD} >= 1.0$. 72

$TSM_{UIP}$    Use isolated Manufacturing in Tactsequence Assignment. $TSM_{UIP} \in \{0, 1\}$ with 0 to disable and 1 to enable. 71

$TSRC$    Tactstation-Requirement Container. 70

$TSS$    Tactsubstation. 67

$TSSC$    Tactstation Container. 67

$TSSRC$    Tactstation-Requirement Container. 70

$TSeq$    Tactsequence. 69

$UKP$    Unbounded Knapsack Problem. 57

$dKP$    d-Dimensional Knapsack Problem. 57

# Nomenclature

**0-1 Knapsack**

Basic form of the Knapsack Problem from which the other variants are derived. 55

**Allocation**

Represents the assignment of each Segment in a Job's Tactsequence to a respective Start-Tact. 97

**Bound**

Consolidation of superfluous branches or problems from a set of problems. 78

**Bounded Knapsack Problem**

Knapsack Problem where item classes contain a bounded number of candiadate items for packing. 57

**Branch-Selection**

Selection of the most suitable branch or problem in a set of problems which will subsequently be broken down into a set of simpler subproblems. 78

**Branching**

Generation of a set of simpler subproblems for a problem selected during Branch-Selection. 78

**Calendar**

Ordered set containing all defined Tacts. 65

**Capacity Supply**

The set of all defined available Processing Capacities. 66

**Change-Making Problem**

Knapsack Problem where the objective is to minimize the number of coins required to achieve a defined total value given an unlimited supply of coins in various denominations. 62

**Collapsing Knapsack Problem**

Knapsack Problem where the addition of an item to a knapsack incurs a per-item weight overhead. 61

**Constraint Container**

Container which contains the MIDs of all Machines which are considered to be constrains for the production system. 104

**Constraint-Factor**

Critical limit for the average relative machine utilization above which machines are considered to be a production system constraint. 101

**Critical Job**

Jobs which must commence production in the Scheduling Horizon in order to be completed by their respective due-dates. 115

**d-Dimensional Knapsack Problem**

Knapsack Problem with multi-dimensional weight. 57

**due-date**

Latest possible date by which the production of a component or Job must be completed. 68

**Electrical Discharge Machining**

A production process used for material removal e.g. during tool manufacturing. The process by achieves material removal from the workpiece by means of repeated eletrical discharges eminating from two electrodes.. 15

**Enterprise Resource Management System**

Software system developed to support and integrate the various business processes such as sales product design and development manufacturing inventory management human resources finance and accounting.. 128

**Equality Knapsack Problem**

Knapsack Problem where the "less than or equal" weight-constraint is replace with a "equals" constraint. 56

**Expanded Scheduling Horizon**

Timespan from which Jobs are chosen for the third phase of Production Scheduling. 122

**Expanding Knapsack Problem**

Knapsack Problem where the addition of an item to a knapsack incurs a per-item weight bonus. 61

**External Process Step**

Process step which is not handled in the context of the synchronized production system. 74

**Fordism**

A production paradigm introduced by HENRY FORD which focuses on a high level of division of labor in conjunction with sationary workers and mobile semi-finished products transported to assembly stations.. 17

**Frozen Zone**

Subinterval within the Scheduling Horizon for which the schedule is considered final. 97

**Idle-Tact**

A Tact in a Sequence Segment which was inserted into a series of process steps during Tactsequence Mapping to render it compatible with a Tacline. 72

**Integer Knapsack Problem with Setup Weights**

Knapsack Problem where addition of an item to knapsack incurs a weight penalty should no item of the same class already be present in the knapsack. 61

**Isolated Manufacturing Step**

Step in a Tactsequence which does not occur in the context of a Tactline and is represented by a single Sequence-Segment. 71

**Job**

Representation of a physical component to be mapped to a Tactsequence and scheduled for production. 68

**Job Admission Time**

Point in time when a Job is added to the Job Pool. 97

**Job Pool**

Set of all defined Jobs to be mapped to Tactsequences and scheduled for production. 68

**Job-Chunk**

Grouping of jobs which share a Rigging-Key and are thus scheduled to arrive a certain Machine for joint processing in a certain Tact. 119

**Job-Chunk Container**

Complete set of Tactstation-Requirements for all Job-Chunks. 119

**Latest Start-Tact**

Latest admissible Tact in which a Job may commence production without violating its Due-Date. 103

**Latest Start-Tact Container**

Container which contains the Latest Start-Tact tuples for all Jobs in the Job Pool. 103

**Lean Management**

Management paradigm derived from the Toyota Production System which today is no longer limited to manufacturing but is applied to other areas such as Innovation Administration and Maintenance.. 19

**Levenshtein Distance**

Minimum number of insertion deletion or substitution operations necessary to transform one string $S_1$ into another string $S_2$. 52

**Logistics Tacts Before Entering Tactline**

Fixed number of empty Tacts inserted into a Tactsequence to allow for the loading of a job onto a Tactpallet. 101

**Logistics Tacts Before Isolated Processing**

Fixed number of empty Tacts inserted into a Tactsequence to allow for the transferring of a job to isolated processing. 101

**Logistics-Tact**

Tacts inserted between the Segments of a Tactsequence to represent logistics operations such as loading a Job onto a Tactpallet or transporting a job to isolated processings. 99

**Machine**

Station performing a manufacturing task which is either a physical machine or a manual workstation. 66

**Machine Complement**

Set of all Machines on the shop-floor. 66

**Maximum Number of Wait-Tact**

Represents the state of scheduling algorithm i.e. all created Allocations at a given time. 101

**maximum-capacity**

Absolute maximum of processing capacity per Tact which can be attained using instruments such as overtime and labor leasing. 66

**Min-Max 0-1 Knapsack Problem**

Knapsack Problem where the objective is to find a combination of items which maximizes the minimum of all solution value vector dimensions. 60

**Minimization Knapsack Problem**

Knapsack Problem where the objective is to minimize the cost subject to a minimal weight constraint. 57

**Multiobjective Knapsack Problem**

Knapsack Problem with multi-dimensional value. 59

**Multiple Choice Knapsack Problem**

Knapsack Problem where multiple items classes exists within which exactly one item must be chosen. 58

**Multiple Knapsack Problem**

Knapsack Problem where the multiple knapsacks each subject to weight constraint must be filled while optimizing overall value. 58

**Precedence Constraint Knapsack Problem**

> Knapsack Problem where the addition of items to knapsack can be contingent upon other items being added to the knapsack first. 59

**Process Step Series**

> Ordered set of the essential sequential manufacturing steps after removal of Virtual and External Process Steps. Each remaining process step includes references to potential target machines and required machine capacities. 69

**Process Step Container**

> Container holding all process steps for all jobs. 69

**Processing Capacity**

> Available target- and maximum processing capacity for Machine during a Tact. 66

**Production Scheduling**

> Assignment of Tactsequences with their respective Sequence-Segments to Tacts for processing. 31

**Production Scheduling Time**

> Point in time when Production Scheduling takes place. 97

**Production Steps**

> Element in routing or Process Step Series which includes target machines and required machine capacities for each step. 69

**Quadratic Knapsack Problem**

> Knapsack Problem where value of an items depends on the other items included in the knapsack. 58

**Relational Algebra**

> Procedural query language introduced by Codd in 1970 consisting of a number mathematical operations that serve to compare modify and combine tuples within different relations. 34

**Relational Calculus**

Declarative query language introduced by Codd in 1972 consisting of a number mathematical operations that serve to compare modify and combine tuples within different relations. 38

**Relational Model**

Model introduced in 1969 by E.F. Codd to structure and store large amounts of data. 33

**Routing**

Ordered set of manufacturing steps created by production engineering which includes Virtual and External Process Steps. 70

**Scenario**

Represents the state of scheduling algorithm i.e. all created Allocations at a given time. 98

**Scheduling End Tact**

Last Tact in the Scheduling Horizon. 97

**Scheduling Horizon**

Timespan for which Production Scheduling is performed. 97, 100

**Scheduling Horizon Expansion Factor**

Factor by which the Scheduling Horizon is expanded to choose additional jobs during the third phase of Production Scheduling. 102

**Scheduling Start Tact**

First Tact in the Scheduling Horizon. 97, 100

**Segment Container**

Full complement of all Sequence Segments for all jobs. 69

**Sequence-Segment**

Series of one or multiple consecutive Tactstations in a Tactsequence through which a bundle of components must pass without pausing or being separated. 30

**Splice-Point**

Points between Segments at which both Logistics- and Wait-Tacts can be inserted when performing Production Scheduling. 99

**Tact**

Defined duration of time during which production is performed. 23

**Tact Horizon**

Interval beginning with the Scheduling Start Tact and ending on the latest possible Tact in which a scheduled Job could require processing. 97

**Tactline**

Series of Tactstations through which a Tactpallet must pass on its path through production facilities. 23, 66

**Tactline Container**

Set containing all defined Tactlines. 66

**Tactpallet**

Logistical instrument used to carry a bundle of components through a series of Tactstations in a Tactline. In the tool manufacturing industry these physical containers are often wooden pallets. 23, 66

**Tactsequence**

Series of Tactstations generated through which a job must pass to complete the manufacturing process. 29

**Tactsequence Dilation**

Effect occurring during Tactsequence Mapping when a Tactline which is used in a Sequence-Segment contains a number of Tactstations which is greater than the number of covered Process Steps. 72

**Tactsequence Mapping**

Generation of the most suitable Tactsequence to represent the manufacturing steps for a job based on a preexisting production system setup and exogenous objectives and restrictions. 29

**Tactstation**

Station within a Tactline to which a Tactpallet is transported between Tacts. 24

**Tactstation Container**

Set containing all defined Tactstations contained in all Tactlines. 67

**Tactstation-Requirement**

Series of Tactstations through which a Tactpallet or - in the case of an Isolated Manufacturing Step - a tool component must pass. Tactstation-Requirements are in turn comprised or zero one or multiple Tactsubstation-Requirements.. 29, 70

**Tactstation-Requirement Container**

Complete set of Tactstation-Requirements for all Jobs. 70

**Tactstations**

Isolated Manufacturing station or station in a Tactline. Tactstations are comprised of one or multiple Tactsubstations. 67

**Tactsubstation**

Element within a Tactstation which maps directly to production machine or workplace on the factory floor. 24, 67

**Tactsubstation-Requirement Container**

Complete set of Tactsubstation-Requirements for all Tactstationrequirments in all Jobs. 70

**Tactsubstation Container**

Set containing all defined Tactsubstations in all Tacstations contained in all Tactlines. 67

**target-capacity**

Ideal amount of processing capacity to be used during one Tact from the company-perspective. 66

**Taylorism**

Set of management principles introduced by FREDERICK WINSLOW TAYLOR which focus on a "scientific" analysis of work in conjunction with a high level of worker specialization and a strong focus on monetary incetivation.. 17

**Toyota Production System**

Poduction system developed by Toyota in post-war Japan. The TPS builds on the ideas pioneered by Ford but focuses on increasing production system flexibility and avoiding waste.. 18

**TSeqCosts**

A tuple which consists of inventory costs per Tact (*ICost*), the logistics cost of loading a component onto a single Tactpallet (*TPLCost*) and finally the logistics costs of transporting a component individually for Isolated Manufacturing (*IPLCost*).. 73

**Unbounded Knapsack Problem**

Knapsack Problem where item classes contain an unbounded number of candiadate items for packing. 57

**Virtual Process Step**

Process step which does not represent a manufacturing step leading to a physical change in the semi-finished product and does not require the product to be physically present for its completion. 74

**Wait-Tact**

Tacts inserted between the Segments of a Tactsequence to represent Tacts in which a job is in storage. A limited number of Wait-Tacts can optionally be inserted during Production Scheduling if it provides a better machine utilization while adhering to the Job's due-date. 98

# Introduction

A cornerstone of companies' success is a strong customer orientation, i.e. identifying customer needs and tailoring products accordingly.[1] In the past decades, a customer demand for individualized products[2] has co-evolved with rapid product renewals. To satisfy heterogeneous market demands companies have increased their range of products by extending product variety and seeking differentiation.[3] To meet the demand for rapid product renewal, product life cycles have shortened or put differently, the frequency of new generation product launches has increased.[4] A greater product variety in combination with short time-to-market necessitates improvements to the process of value creation to remain competitive[5].

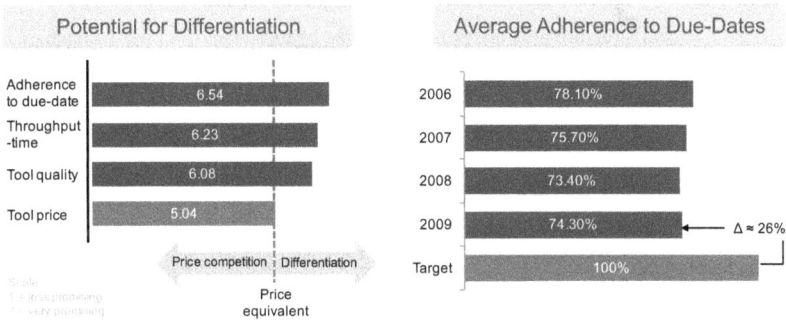

Figure 1.1: Results of a study among tool manufacturers

Tool manufacturers aim to provide producing companies with the tools and machinery mandatory to perform series production. Thus tool manufacturing companies occupy a key position in the value creation process. Consequently, the aforementioned customer trends

---

[1] Cf. WAGNER (2004) Kundenorientierung: Der Königsweg zum Unternehmenserfolg, p. 2.

[2] Cf. MOSER (2007) Mass customization strategies: Development of a competence-based framework for identifying different mass customization strategies, p. 1.

[3] Cf. SCHUH (2012) Handbuch Produktion und Management 3, pp. a; BER (2012) Mastering product complexity, pp. b.

[4] Cf. ABELE AND REINHART (2011) Zukunft der Produktion: Herausforderungen, Forschungsfelder, Chancen, pp. x; SPATH et al. (2001) Vom Markt zum Markt: Produktentstehung als zyklischer Prozess, pp. y.

[5] Cf. FELDHUSEN AND GROTE (2013) Pahl/Beitz Konstruktionslehre: Methoden und Anwendung erfolgreicher Produktentwicklung, pp. z; RISSE (2003) Time-to-Market-Management in der Automobilindustrie: Ein Gestaltungsrahmen für ein logistikorientiertes Anlaufmanagement, pp. 4.

also affect the tool manufacturing industry. While a greater product variety accounts for an increase in the demand for tools, shorter product life cycles entail a higher time-pressure for tool manufacturers. Additionally, competition and price-pressure have grown with the market entry of new tool manufacturers from Asia.

German tool manufacturers historically have attempted to offset their high labor-costs by investing heavily into process automation and by providing products with a lower total cost of ownership than their international competitors.[6] Today, however, technologically superior tools and a high degree of automation are insufficient to retain a strong position in the market.[7] To overcome the challenges illustrated above, new measures to strengthen competitiveness have to be taken.

Machine tools are generally highly specific to their purpose and thus considered to be make-to-order products. A demand for high flexibility and small batch sizes has resulted in a predominance of organizational structures similar to those found in handicraft businesses.[8] This organizational structure also arises from the size of tool manufacturing companies in Germany, which are mostly considered to be small or medium enterprises. In contrast to the requirement of flexibility, highly varied, complex sequences of production steps lead to poor transparency and efficiency of the overall job shop production process. Processing-time in tool manufacturing often amounts to less than 5% of tools' total throughput-time.[9] Improvements to these performance indicators are, however, non-trivial owing to the aforementioned organizational structures.

In fact, a study conducted among tool manufacturers by the Laboratory for Machine Tools and Production Engineering of the RWTH Aachen reveals that adherence to due-dates and throughput-time are considered to be the two leading differentiators (refer to left part of Figure 1.1)[10]. Thus the study substantiates the conclusions drawn from the analysis of the trends described in the first paragraph. Furthermore, respondents indicated an average of 74.3% adherence to due-dates. In light of the fact that the process steps following tool manufacturing in the value chain often produce "just-in-time", the necessity for a disruptive improvement of tool manufacturing processes becomes clear.

[6]  Cf. BOOS (2008) Methodik zur Gestaltung und Bewertung von modularen Werkzeugen, pp. 151.
[7]  Cf. ZISKOVEN (2013) Methodik zur Gestaltung und Auftragseinplanung einer getakteten Fertigung im Werkzeugbau, p. 2.
[8]  Cf. PITSCH et al. (2015) Getaktete Fertigung im Werkzeugbau, p. 8.
[9]  Cf. Ibid., p. 9.
[10]  Cf. (2010)Operative Exzellenz im Werkzeug- und Formenbau, p. 16.

In an effort to boost competitiveness, many companies in the tool manufacturing industry have begun industrializing their processes and organizational structures. Some companies have begun focusing on their core competencies - resulting in out-sourcing production or components - others have focused on standardizing products and processes, implementing flow production or even experimenting with process synchronization.[11] Especially the latter has shown very good results in the few cases in which it has been implemented. Reductions of throughput-time by up to 60% and improvements of the adherence to due-dates by up to 70% have been observed[12].

While the prior work by ZISKOVEN and ZWANZIG has been instrumental to allowing several tool manufacturers to transition to a synchronized manufacturing system, significant short-term manual planning effort remains, which renders daily operations costly and complex. Thus to take full advantage of the potentials inherent in a synchronized flow of materials on the factory floor, the following research question needs to be addressed:

*What activities are necessary to support the short-term planning of synchronized tool manufacturing, which takes due-dates into account, minimizes throughput-time and makes efficient use of machinery?*

In the following Sub-Chapters the methodology guiding the research in this thesis is outlined (Chapter 1.1) and the resulting thesis structure is presented (Chapter 1.2).

## Research Methodology

Prior to discussing the structure of this thesis, this Chapter will briefly outline the underlying research methodology. Central determinant of the research process is its classification in HILL's and ULRICH's spectrum of the sciences shown in Figure 1.2.[13] Generally, the sciences are divided into formal and empirical sciences. Formal science focuses on analytical statements and refers to abstract objects, which do not exist in the real world, e.g. numbers, symbols and sets.The objective is the construction of systems with a set of rules, which can only be checked for correctness using logic.[14]

---

[11] Cf. KLOTZBACH (2007) Gestaltungsmodell für den industriellen Werkzeugbau, pp. 239.
[12] Cf. WILLRETT (2011) Industrieanzeiger, p. 31.
[13] Cf. ULRICH (1982) Die Unternehmung, p. 305.
[14] Cf. SCHANZ Wissenschaftstheoretische Grundfragen der Führungsforschung, p. 2192.

```
                          ┌──────────── Science ────────────┐
                          ▼                                  ▼
              Formal Science                            Empirical Science
                          │                           ┌──────────┴──────────┐
                          │                           ▼                     ▼
                          │                  Fundamental Science    Applied Action Science
                          ▼                           ▼                     ▼
              ┌────────────────────┐        ┌────────────────────┐  ┌────────────────────┐
              │   Construction of  │        │ Explanation of Empirical │  │  Analysis of Human │
              │  Character Systems │        │   Reality Extract   │  │ Action Alternatives│
              └────────────────────┘        └────────────────────┘  └────────────────────┘
                          │                           └──── Engineering science ────┘
          Philisophy, Logic, Mathematics       Natural science         Social science
```

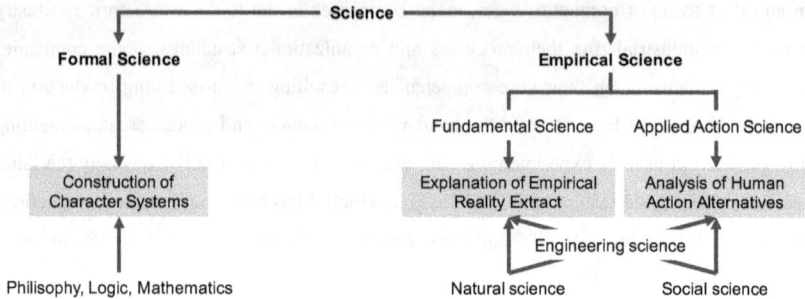

Figure 1.2: Classification of the sciences

Empirical Sciences, on the other hand, are synthetic in nature[15] and aim to describe, explain and offer guidelines for the design of extracts of the observable world. Empirical science can again be subdivided into fundamental and applied action sciences. While former focus on establishing a general understanding of nature, the latter serve to explain the actions of individuals of societies[16]. Engineering science contains elements of both and is thus positioned between fundamental and applied action sciences[17]. Within Engineering science the original problem statement underlying this thesis can be attributed to applied action science. During the search for solutions, however, the research process will also draw upon elements generated in the natural and formal sciences.

While science is the composition of facts, theories and methods, research methodology describes the path taken to derive the aforementioned. Research is the process with which the elements of science are compiled to form a continuously growing portfolio, which constitutes the scientific state of the art and experience[18]. The problems representing the origins from which research begins are generally very different for the different branches of science described above. In the case of formal sciences the theme guiding the definition of research foci are incomplete areas of existing logical constructs. In fundamental science the basis is usually a discrepancy between theory and empirical observations.[19] In the applied action sciences, however, the objective is the rather pragmatic generation of knowledge yielding

---

[15] Cf. Ibid..
[16] Cf. Ibid..
[17] Cf. HILL AND ULRICH Wissenschaftliche Aspekte ausgewählter betriebswirtschaftlicher Konzeptionen, p. 163.
[18] Cf. KUHN (2001) Die Struktur wissenschaftlicher Revolutionen, p. 18.
[19] Cf. ULRICH (1982) Die Unternehmung, pp. 173.

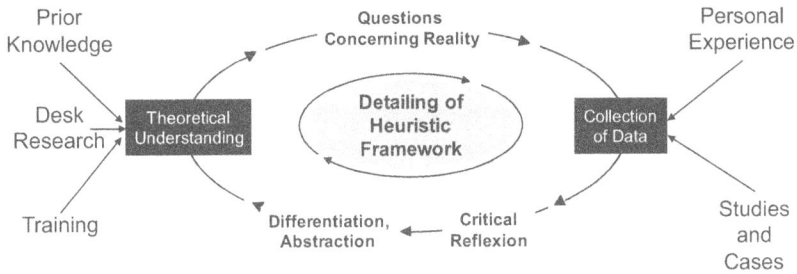

Figure 1.3: Cycle of research according to KUBICEK

a practical use[20]. Thus the selection of research questions focuses on fields, within which insufficient knowledge exists to solve a practical problem.

The usage of the purely deductive, empirical strategy used in fundamental science[21] is not instrumental in the context of applied action science and is in some cases even impedimental[22]. This line of argumentation leads to a different research methodology in the applied action sciences, which is referred to as explorative research[23]. The underlying idea is not to verify hypotheses derived from theory in reality, but rather to develop scientific statements to generate new realities[24].

The construction of scientific statements is performed using an iterative process, which is guided by theoretical objectives and is based on systematically obtained experience[25]. This process thus focuses both on the acquisition of experience and the creative generation of new theoretical statements[26]. The superordinate objective is to generate new insight into reality. To this end theoretical questions regarding reality are examined to generate experience. This experience is then used as a basis for the derivation of new questions.[27] The resulting clycle is shown in Figure 1.3.

ULRICH's process of applied research, shown in Figure 1.4, is based on KUBICEK's work,

---

[20] Cf. KUBICEK (1976) Heuristische Bezugsrahmen und heuristisch angelegte Forschungsdesign als Elemente einer Konstruktionsstrategie empirischer Forschung, pp. 5.
[21] Cf. HEMPEL AND OPPENHEIM (1948) Philosophy of Science, pp. 135.
[22] Cf. KUBICEK (1976) Heuristische Bezugsrahmen und heuristisch angelegte Forschungsdesign als Elemente einer Konstruktionsstrategie empirischer Forschung, pp. 5.
[23] Cf. Ibid., p. 7.
[24] Cf. KROMREY (1991) Empirische Sozialforschung, p. 20.
[25] Cf. KUBICEK (1976) Heuristische Bezugsrahmen und heuristisch angelegte Forschungsdesign als Elemente einer Konstruktionsstrategie empirischer Forschung, p. 13.
[26] Cf. MALIK (1992) Strategie des Managements komplexer Systeme: Ein Beitrag zur Management-Kybernetik evolutionärer Systeme, pp. 255.
[27] Cf. TOMCZAK (1992) Marketing ZFP, p. 84.

but places a stronger emphasis on the actual application of the obtained insights. Therefore ULRICH's process continues after the evaluation of developed solutions and demands a prototypical evaluation and subsequent implementation in practice.[28] This process guided the work underlying this thesis and after completion of the underlying research all contained steps executed. As the culmination of the research underlying this thesis, the developed algorithms were implemented into a software-product, which now supports the daily production planning activities of three major German automotive first-tier suppliers.

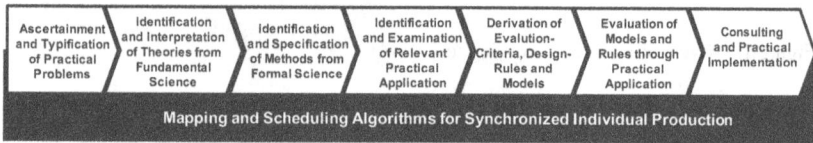

Figure 1.4: Process of applied research according to ULRICH

# Thesis Structure

As noted in the previous Chapter, the research underlying this thesis was guided by the process of applied research and thus ULRICH's work has a strong influence on the structure of this resulting thesis. Chapter 1 serves to introduce the reader to the subject-matter, gives an overview of problems resulting for industrial practice, presents the research question and thereby provides an outline of the intended results.

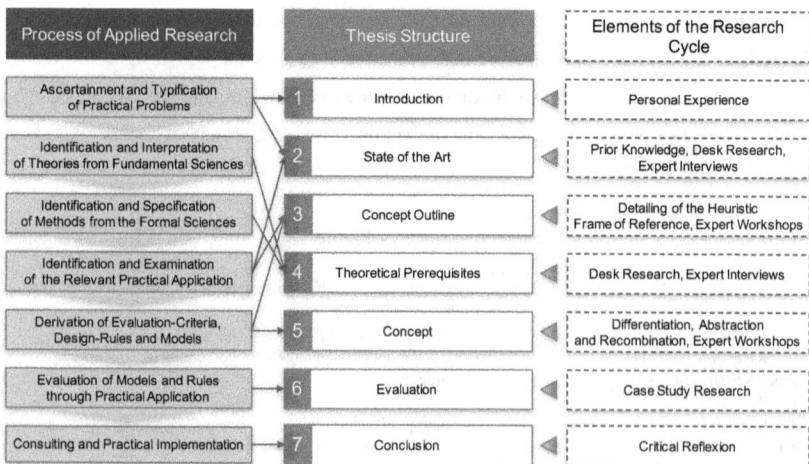

Figure 1.5: Thesis structure

[28] Cf. ULRICH (1982) Die Unternehmung, pp. 192.

Chapter 2 describes the thesis' frame of reference by giving an in-depth overview of tools and tool manufacturing. The third Sub-Chapter introduces the concept of industrialization. The level of detail in this chapter is iteratively increased by elaborating on relevant elements discovered on the previous level of detail. Accordingly, in a first step, industrialization and its effect on industry and society are discussed. In a second iteration the application of industrialization to the tool manufacturing industry are examined. In a third and final iteration, the subject of synchronization within industrialized tool manufacturing is discussed. On this level of detail two research gaps are identified from which two research questions are then derived.

Chapter 3 presents an outline of the concept, which gives an overview of the two short-term planning activities, i.e. "Tactsequence Mapping" and "Production Scheduling", which are necessary to allow companies to take advantage of a synchronized production system. This Chapter concludes with a brief introduction of the notation used in later Chapters to describe algorithms, which can automate aforementioned activities.

In Chapter 4 necessary theoretical fundamentals from the formal and fundamental sciences are introduced. First this Chapter provides a background in the area of combinatorial optimization and mathematical complexity theory. Subsequently, two specific problems from the areas of operations research and computer science are presented, which will be used as basic tools to construct the algorithms in Chapter 5.

Chapter 5 provides a detailed description of the algorithms designed to solve the "Tactsequence Mapping" and "Production Scheduling" using the notation introduced with the concept outline. Aside from theoretical requirements arising from the problem definition, the algorithm design takes requirements from industrial practice into account. Chapter 6.1 describes the steps taken to validate the proposed algorithms. In line with KUBICEK's research cycle, the evaluation was performed continuously and thus guided the entire research process. The combination of both a theoretical and practical evaluation is reflected by the two corresponding Sub-Chapters.

Chapter 7 finally summarizes the results generated in the preceding Chapters and provides an outlook on objects of potential future research.

# Chapter 2

# State of the Art

This Chapter provides a detailed description of the thesis' frame of reference by providing an in-depth overview tools, the tool manufacturing sector and finally industrialization.

## Tools

As the name implies, companies in the manufacturing sector aim to provide customers with physical products meeting defined requirements in a timely fashion. In accordance with GUTENBERG's definition, the input necessary to drive the transformation process are referred to as production factors[29]. These are divided into dispositive and elementary factors. The former encompasses all supporting tasks and activities necessary to combine the elementary factors and thus produce a product. Typically dispositive factors include organizational and managerial tasks[30]. Elementary production factors fall into the three categories (human) labor, operating resources and materials[31]. Operating resources are once again subdivided into buildings, machinery, tools and office space[32].

According to the VDI guideline 2815[33] tools are manufacturing equipment used to influence the form or substance of material during the manufacturing process by means of either physical or physical-chemical interactions[34]. Since both hand tools such as hammers, files, screwdrivers etc. and standardized tools such as milling heads, drills, chisels etc. are included in the aforementioned definition, this definition does not apply to this thesis[35]. The German Engineering Federation (VDMA) defines tools as equipment belonging to the following

---

[29] Cf. GUTENBERG (1960) Grundlagen der Betriebswirtschaftslehre, p. 3.
[30] Cf. Ibid., p. 6.
[31] Cf. DYCKHOFF (1995) Grundzüge der Produktionswirtschaft, p. 343.
[32] Cf. BRANKAMP Organisation des Betriebsmittelbaus, p. 13.
[33] Although the guideline was withdrawn in 2001 without replacement, the definition of tools contained herein is still in widespread use today.
[34] Cf. VDI (1978) Begriffe für die Produktionsplanung und -steuerung, p. 2.
[35] Cf. EVERSHEIM AND KLOCKE (1998) Werkzeugbau mit Zukunft: Strategie und Technologie, p. 25.

| Tools | | Jigs | Test Equipment | Models |
|---|---|---|---|---|
| Sheet Metal Processing Tools | Plastic Processing Tools | Processing Jigs | Calipers | Casting Models |
| • Drawing<br>• Cutting<br>• Forming<br>• Bending | • Injection molding<br>• RIM<br>• SMC<br>• Elastomer | • Drills<br>• Mills | Stencils | Design Models |
| Mass Forming Tools | Die Casting Tools | Assembly Jigs | Special-Purpose Machines | |
| Other Tools | | | | Functional Models |

| Other services | | | | |
|---|---|---|---|---|
| Consulting | Prototypes | Delivery of Systems | Repair / Maintenance | Component Manufacturing |

Figure 2.1: Tool Manufacturing - products and services

categories: production measurement and inspection technology, clamping tools, stamping tools, fixtures and molds, cutting tools[36]. Since this definition, however, covers equipment not typically in the spectrum of products produced by the tool manufacturing sector it too is not suitable.

In line with EVERSHEIM ET AL. tools are henceforth considered to be the products of the tool manufacturing sector. Thus this definition encompasses all tools for injection molding, die casting, pressing, stamping and molding (see also Figure 2.1)[37]. A characteristic shared by all of the aforementioned tools is the complete or partial correspondence of tool to workpiece geometry[38].

## Tool Manufacturing

Tool manufacturing can be viewed from an organizational and a process perspective. In the context of the value creation process, tool manufacturing occupies a key [39] position between product development and series production (see Figure 2.2[40]). Thus it greatly influences

---

[36] Cf. SPENNEMANN (2001) Gestaltung von Organisationsstrukturen im Werkzeugbau, p. 11.
[37] Cf. FRICK (2006) Erfolgreiche Geschäftsmodelle im Werkzeugbau, p. 25.
[38] Cf. SCHRÖDER (2003) Aufbau hierarchiearmer Produktionsnetzwerke: Technologiestrategische Option und organisatorische Gestaltungsaufgabe, p. 146.
[39] Cf. MENEZES (2004) European Mold Making, p. 3.
[40] Cf. FRICK (2006) Erfolgreiche Geschäftsmodelle im Werkzeugbau, p. 182.

the timing of market entry and the quality of the ensuing manufacturing process[41] - two major competitive factors for companies in the manufacturing sector[42]. According to SCHUH investment and maintenance costs for production tools account for up 30% of production costs[43]. Embedded between product development and series production, tool manufacturing serves as a supplier to both. Product development can draw upon the know-how regarding product design suitable for efficient production, while series production is not only supplied with operating resources but can also benefit from the experience in setting manufacturing processes (e.g. choosing suitable temperatures, forces or throughput-time parameters). Thus tool manufacturing is in the position to provide knowledge-intensive services to the adjacent process steps.

From an organizational perspective tool manufacturing can either be embedded into a larger organization or function as an independent entity[44]. In the former case, hence referred to as internal tool manufacturing, the organizational unit is integrated directly into the organizational structure of the parent-company[45]. In this case the unit provides tools tailored to the requirements of the parent-company's manufacturing process in exchange for resources. Key advantages hereof lie in improved availability of production tools and the protection of parent-company know-how.[46] When independent legal entities provide production services on the market, hence referred to as external tool manufacturing, these strive to maximize their respective profits. Due to their interaction with a wide range of customers external tool manufacturers can accrue broad customer-independent expertise and pursue differentiation strategy by focusing on specific types of tools. [47]

## Sector Structure

Tool manufacturing in Germany is highly fragmented and comprised of many small and medium enterprises, some of which feature traditional organizational structures similar to those found in handicraft businesses[48]. According to the VDMA classification, tool

---

[41] Cf. HOLMES, RUTHERFORD AND FITZGIBBON Innovation in the Automotive Tool, Die and Mold Industry: A Case Study of the Windsor-Essex Region, p. 125; DÖRING (2010) Konfliktmanagement in der technischen Auftragsabwicklung im Werkzeugbau, p. 31; GIEHLER (2010) Erhöhung der Planungsproduktivität am Beispiel der Auftragsabwicklung im Werkzeugbau, p. 27.
[42] Cf. DI BENEDETTO, C. ANTHONY (1999) Key Success Factors in New Product Launch, p. 539; COOPER (1979) Dimensions of Industrial New Product Success and Failure, p. 100.
[43] Cf. (2010)Operative Exzellenz im Werkzeug- und Formenbau, p. 13.
[44] Cf. EVERSHEIM AND KLOCKE (1998) Werkzeugbau mit Zukunft: Strategie und Technologie, p. 74.
[45] Cf. Ibid., p. 3.
[46] Cf. FRICK (2006) Erfolgreiche Geschäftsmodelle im Werkzeugbau, p. 23.
[47] Cf. SCHLEYER (2006) Erfolgreiches Kooperationsmanagement im Werkzeugbau, p. 22; WESTEKEMPER (2002) Methodik zur Angebotspreisbildung, p. 7.
[48] Cf. BISPING (2003) Integrierte Produktstrukturen für den Werkzeugbau, p. 3.

Tool Manufacturing

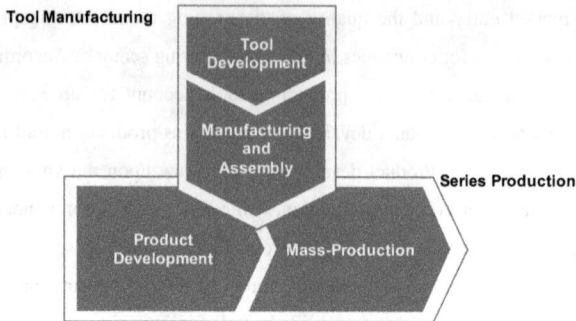

Figure 2.2: Position of tool manufacturing in the chain of value creation

manufacturing as defined in this thesis falls into the precision tool manufacturing sector in which approximately 5000 companies generate a revenue of 9,3 billion Euro relying on approx. 70,000 employees[49]. Due to the prevalence of internal tool manufacturing and the fact that only a small percentage of tool manufacturers are members of a tool manufacturers organization, an exact census of tool manufacturers is not possible[50]. The number of large companies with more than 300 employees is, however, exceedingly small[51]. While this distribution also applies to most other industrialized nations, low wage countries represent exceptions to this rule. In China more than 80 percent of tool manufacturing companies have more than 100 employees and more than 14 percent have more than 500 employees[52].

Gross revenue for the tool manufacturing sector in Germany has seen continuous year over year increases in the past fifteen years. The only interruption of this trend, the economic turbulences in the year 2009, caused a 26% drop in revenue shown in Figure 2.3[53]. The resulting abundance of manufacturing capacities on the market led to a drop in tool prices of 12,3%[54]. Due to strong exports the German tool manufacturing sector, however, rebounded and surpassed the previous record revenues within two years.

Despite the aforementioned positive economic developments in the previous years international competition remains strong and the market is characterized by an intensifying price war. Strong competition with highly diversified technological expertise can be found in many countries such as the United States, China, Japan, Italy, Korea, Spain, Portugal, Great

---

[49] Cf. VDMA (2014) Statistisches Handbuch für den Machinenbau: Ausgabe 2014, p. 120.
[50] Cf. MENEZES (2004) European Mold Making, p. 3.
[51] Cf. FRICK (2006) Erfolgreiche Geschäftsmodelle im Werkzeugbau, pp. 27.
[52] Cf. BOOS et al. (2015) Tooling in China, p. 10.
[53] Cf. VDMA (2014) Statistisches Handbuch für den Machinenbau: Ausgabe 2014, p. 120.
[54] Cf. (2010)Operative Exzellenz im Werkzeug- und Formenbau, pp. 13-14.

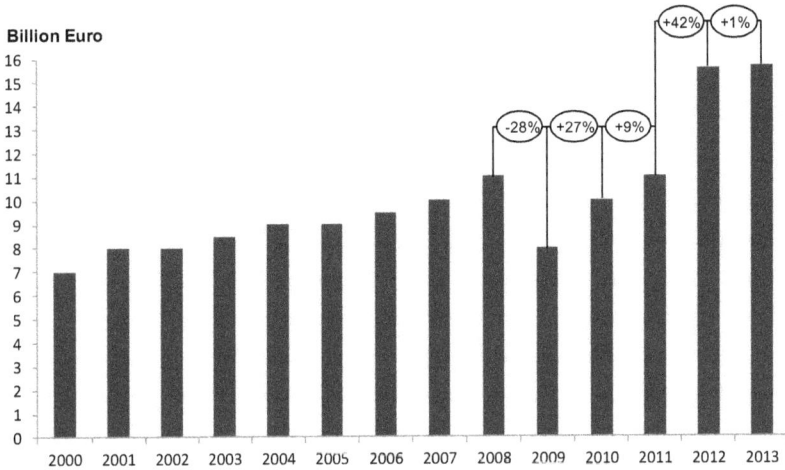

Figure 2.3: Development of revenues in the precision tool manufacturing sector

Britain, Malaysia, Belgium, South Africa, Switzerland and Brazil[55]. Due to a differentiation strategy of focusing on high-value tools and above average productivity per employee, the average revenue per employee in Germany is the highest in the world[56][57].

## Process Chain

As described in Chapter 2.2 tool manufacturing is positioned between the product development and series production in the value creation process of a manufacturing company. To ensure economical order fulfillment simultaneous engineering is often used to organize the necessary process steps from order acquisition, tool development, production engineering, production and assembly to the final tool handover[58]. Based on the FRICK'S prototypical process chain for the tool manufacturing sector Figure 2.4[59] provides an overview hereof. Depending on the strategic orientation of a company all or only some of the aforementioned steps are performed [60]. This Chapter provides a brief description of the five major process steps and the activities contained therein.

---

[55] Cf. MENEZES (2004) European Mold Making, p. 4; KOZIELSKI (2010) Integratives Kennzahlensystem für den Werkzeugbau, p. 17.
[56] Cf. SCHUH AND KLOTZBACH (29.11.2005) Situation of the Moldmaking Industry and Success Factors for its Development.
[57] Cf. FRICK (2006) Erfolgreiche Geschäftsmodelle im Werkzeugbau, pp. 24-25.
[58] Cf. WESTFECHTEL (1999) Models and Tools for Managing Development Processes, p. 17.
[59] Cf. FRICK (2006) Erfolgreiche Geschäftsmodelle im Werkzeugbau, p. 27.
[60] Cf. ZWANZIG (2010) Taktung der Unikatfertigung am Beispiel des Werkzeugbaus, p. 28.

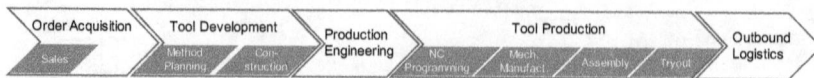

Figure 2.4: Prototypical tool manufacturing process chain

Order acquisition, the initial step in the process chain, contains both sales and project management activities. As the first point of customer contact sales activities provide customers with offers tailored to their specific requirements[61]. Customer requirements are first structured and then compiled into a performance specification from which an initial project sketch is drafted[62]. In this context the calculation of resulting tool costs is a challenging activity. Due to the one-off nature of tool manufacturing projects and incomplete information provided by product development, estimates are often rough. Studies show that with identical input-data estimates of final tools costs in this phase can differ from each other by more than 100 percent[63], while the difference between estimated and actual tool costs can amount to 70 percent[64]. Some tool manufacturers attempt to improve estimation precision by basing estimates on similar projects previously completed[65]. Due to the high degree of individualization project management activities are initiated in Sales [66] and continue until a tool has been successfully delivered to the customer. Project management encompasses all management activities necessary to achieve a well-defined project objective in a defined manner, time and with resources[67]. While frequent customer induced changes and tight time constraints highlight the importance of these activities, systematic project management often does not find application in industrial practice[68].

Following order acquisition, the tool development which is in turn comprised of method planning and tool construction, ensues. Method planning entails definition of methods and processes for the production of the final product[69]. Thus this step of the tool manufacturing value chain is contained in the production engineering step of the superordinate value chain of the final product. During method planning the technological tool concept is defined. This includes the tool structure, number of components, tool materials and the number of forming

[61] Cf. KLOCKE AND SCHUH (2005) Zukunftsstudie Werkzeug- und Formenbau, p. 53.

[62] Cf. ZWANZIG (2010) Taktung der Unikatfertigung am Beispiel des Werkzeugbaus, p. 28.

[63] Cf. SCHLÜTER (23.8.2006) Vorkalkulationsmethoden von Spritzgießwerkzeugen: Auf Basis der Ähnlichkeitsbetrachtung, p. 1.

[64] Cf. EVERSHEIM AND KLOCKE (1998) Werkzeugbau mit Zukunft: Strategie und Technologie, p. 45.

[65] Cf. BOOS (2008) Methodik zur Gestaltung und Bewertung von modularen Werkzeugen, p. 42.

[66] Cf. SCHLEYER (2006) Erfolgreiches Kooperationsmanagement im Werkzeugbau, p. 30.

[67] Cf. KESSLER AND WINKELHOFER (2004) Projektmanagement: Leitfaden zur Steuerung und Führung von Projekten ; mit 42 Tabellen, p. 10.

[68] Cf. BOOS (2008) Methodik zur Gestaltung und Bewertung von modularen Werkzeugen, p. 42.

[69] Cf. NEDESS AND HAUER (1997) Organisation des Produktionsprozesses, p. 228.

stages. From a technological perspective, method planning and tool construction are the stages in the development process which require the greatest amount of technological know-how.[70]

Production engineering aims to devise a suitable production process for the fully constructed tool it receives from the product development step. Devising a suitable production process entails determining the necessary technological sequence of manufacturing operations, defining the semi-finished products for the various stages and defining due-dates for the components comprising the developed tool.[71]

The tool production step consists of four major sub-processes: NC programming, mechanical production, assembly and try-out. Depending on the reference process definition, NC-programming can also be regarded as being a part of production engineering[72]. NC programming is the creation of the programs to control the machine tools which perform the mechanical processing of tool components and *Electrical Discharge Machining* ($EDM$) electrodes[73]. Since the design of the NC programs is in essence a non-physical production step usually performed on the factory floor and thus requiring suitable scheduling, FRICK's definition presented above represents a better fit in the context of this thesis. Tools mainly consist of cast pieces of metal, which are processed using a series of milling, turning, drilling, polishing and EDM operations[74]. Following mechanical production, the individual components are assembled yielding a complete machine tool[75]. This step is to date primarily performed by skilled manual labor.[76] The tryout finally ensures that the machine tool can be used in a capable process to produce high quality products[77].

Outbound logistics marks the final step of the tool manufacturing process and serves to initiate the logistical operations necessary to deliver the finished tool to the customer.

# Industrialization and Synchronization

This Chapter examines the implications of industrialization for tools and the tool manufacturing industry. The level of detail in this chapter is iteratively increased by elaborating on relevant elements discovered on the previous level of detail. Accordingly

---

[70] Cf. SCHLEYER (2006) Erfolgreiches Kooperationsmanagement im Werkzeugbau, p. 31.
[71] Cf. RITZ (1999) Bewertung technischer Änderungen im Werkzeugbau, p. 9.
[72] Cf. KLOTZBACH (2007) Gestaltungsmodell für den industriellen Werkzeugbaus, p. 49.
[73] Cf. ZWANZIG (2010) Taktung der Unikatfertigung am Beispiel des Werkzeugbaus, p. 30.
[74] Cf. FRICK (2006) Erfolgreiche Geschäftsmodelle im Werkzeugbau, p. 28.
[75] Cf. ZWANZIG (2010) Taktung der Unikatfertigung am Beispiel des Werkzeugbaus, p. 30.
[76] Cf. RITZ (1999) Bewertung technischer Änderungen im Werkzeugbau, p. 9.
[77] Cf. FRICKER (2005) Strategische Stringenz im Werkzeug- und Formenbau, p. 47.

in a first step, industrialization and its effect on industry and society are discussed. In a second iteration the application of industrialization to the tool manufacturing industry are examined. In a third and final iteration, the subject of synchronization within industrialized tool manufacturing is discussed.

## Historical Context

The Industrial Revolution, which was sparked in England and then spread to continental Europe, the United States and finally Asia set in motion a shift from an agrarian to an industrial society. Since this revolution began in 1760 and lasted for approximately 80 years, it should be thought of not as one single transformative incident but rather a series of events delivering incremental change[78]. Over the years this has led to a genesis of many conflicting definitions of the term Industrial Revolution[79]. For the purposes of this thesis, the characterization of the Industrial Revolution as a period during which the proportion of industrialized labor increased relative to agrarian and service labor can be considered valid[80]. From the perspective of economical development it represents a pivotal change, which represents the basis of contemporary economical and political systems[81].

Prior to the Industrial Revolution the production of goods was generally decentralized and carried out by craftsmen. One example of this was the publishing industry in the early 18th century where publishers purchased the necessary raw materials. These were then given to a craftsmen, who produced cloth in their workshops which were often part of their homes. The finished product was then again distributed by the publisher. This system is characterized by the separation of capital and the means of production.[82] A first intermediary step on the path to the modern factory was the so-called "manufactory", where the centralized manual production of goods on publishers' machines was performed by wageworkers.[83] This represented a major change in the production paradigm: the investment risk formerly born by the craftsman was now shifted to the owner of production facility. Thus the motivation to ensure production remained competitive shifted to the entrepreneur. Toward the end to the 19th century the term "factory" was used to describe production facilities, where powered machines performed work formerly carried out manually.[84]

---

[78] Cf. MOKYR (1999) The British industrial revolution: An economic perspective / edited by Joel Mokyr, pp. 2.
[79] Cf. PAULINYI Die Entstehung des Fabriksystems in Großbrittannien, p. 12.
[80] Cf. Ibid., p. 9.
[81] Cf. (1980)Geschichte der Arbeit: Vom Alten Ägypten bis zur Gegenwart, p. 193.
[82] Cf. Ibid., pp. 197.
[83] Cf. Ibid., p. 204.
[84] Cf. Ibid., p. 44.

In 1911 the American engineer FREDERICK WINSLOW TAYLOR published his seminal book "The Principles of Scientific Management", in which he presented a management theory today often referred to as *Taylorism*. In this book be describes four basic principles:

1. The first principle states that it is the management's task to analyze and comprehend, decompose and describe each step comprising the work performed on the factory floor. Each action on the factory floor must be enabled by an action on the management level.[85]

2. The second principle demands that management is to select and train workers specifically in accordance with the tasks they are to perform. Thus management must understand the physical and intellectual capabilities of each worker, choose a suitable task and divide labor accordingly[86]

3. The third principle underlines the importance of the mutual agreement of management and worker and thus the alignment of the objectives for both parties. The favored way of achieving this is the payment of salaries dependent on worker productivity. This helps to ensure that defined rules are interpreted in the spirit they were created. Furthermore, workers are encouraged to provide input regarding the improvement of the production process.[87]

4. The fourth and final principle applies to the distribution of responsibilities between management and representatives of the workforce. This serves the objective of fostering cooperation between both parties and thus avoiding strife and strikes.

TAYLOR's principles are based on the supposition that humans primarily seek happiness in wealth and the associated ability to consume. Thus the fair distribution of profit in accordance with performance serves to motivate workers to work with maximum productivity and efficiency.[88] A negative factor often associated with Scientific Management are the repetitive character and lack of creative freedom associated with highly specialized work. Nevertheless TAYLOR's principles have a strong influence on work to this day.

In 1913 HENRY FORD the Chairman of Ford Motor Company at that time, expanded on TAYLOR's ideas and introduced a production paradigm referred to as *Fordism* or assembly line production. FORD's production system was characterized by three major aspects[89]:

---

[85] Cf. TAYLOR (1919) Die Grundsätze wissenschaftlicher Betriebsführung, p. 38.
[86] Cf. Ibid..
[87] Cf. Ibid..
[88] Cf. HOYER AND KNUTH (1976) Die teilautonome Gruppe. Strategie des Kapitals oder Chance für die Arbeiter?, p. 150.
[89] Cf. FORD AND THESING (1923) Mein Leben und Werk, pp. 93.

1. Workers and tools are positioned in accordance with the sequence of necessary assembly steps. This serves to minimize the distance traveled by the semi-finished product.

2. Workers remain stationary while transport systems deliver semi-finished products and necessary materials to their station.

3. Work was divided into small portions, which could be completed in short periods of time. This accommodates the continuous flow of materials from one assembly station to the next.

The extensive division of labor enabled FORD to make efficient use of unskilled labor[90] to perform assembly tasks and allowed Engineers to concentrate on the continuous optimization of the product and production system. This proved successful as prices for the Model T[91] dropped from 1200$ in 1908 to 259$ in the final production year of 1927[92]. The impressive improvement to production efficiency was made possible by standardization: by 1914 all color options were dropped and all Model Ts were painted black.[93] For many years this production principle was deemed primarily suited to mass production and a poor fit for the production of products with many variants and small lot sizes[94].

Due to the high number of product variants necessary to satisfy the Japanese market, the **Toyota Production System** was developed in the second half of the 20th century to increase the flexibility of mass production[95]. Compared to the mass production of vehicles based on projected sales popular in the United States, Toyota aimed to reduce the throughput-time from order to delivery and attempt to base future production on orders which had been already placed.[96] This allowed for significant reductions of the number of semi-finished products in production facilities and thus for reduced inventory costs. In general, Toyota strived to avoid unnecessary costs and waste[97] in his company. To satisfy the large number of different vehicle models, Toyota furthermore designed its assembly lines to enable the production of multiple different models on one line.

---

[90] Cf. CLARKE (2002) Forms and functions of standardisation in production systems of the automotive industry: The case of Mercedes-Benz: Dissertation Freie Universität Berlin, p. 27.
[91] The Model T was the first vehicle manufactured based on the aforementioned principles.
[92] Cf. FORD AND THESING (1930) Henry Ford und trotzdem Vorwärts!, pp. 189.
[93] Cf. JAYARĀMAN (2007) Management Icons, p. 12.
[94] Cf. WOMACK, JONES AND ROOS (1991) Die zweite Revolution in der Autoindustrie: Konsequenzen aus der weltweiten Studie aus dem Massachusetts Institute of Technology, p. 19.
[95] Cf. ŌNO (1993) Das Toyota-Produktionssystem, p. 22.
[96] Cf. ULRICH (1982) Die Unternehmung, p. 100.
[97] The Toyota Production System specifically addressed seven different types of waste (muda). For more information on the subject please refer to ŌNO (1993) Das Toyota-Produktionssystem, 47.

Increasing product variety and decreasing lot sizes began to cause problems for the automotive industry in the United States in the 1980s. In a large international study conducted by the the Massachusetts Institute of Technology WOMACK, JONES and ROOS examined the production systems of international car manufacturers. The results revealed that Japanese manufacturers were considerably more agile and required considerably fewer employees in both development and production to perform similar tasks[98]. Originally coined by KRAFCIK in his 1988 paper "Triumph of the Lean Production System" the term "Lean Production" was further popularized by inclusion in the study results published in the book "The machine that changed the world". [99]. In the following years the ideas underlying the Toyota Production System were generalized, abstracted and synthesized into the *Lean Management* principles. Lean Management included a wide array of supporting tools and methods such as the Kanban-System[100], the Single-Minute-Exchange-of-Die[101], Mixed Model Assembly Lines, Value Stream Mapping[102] and many more. Today Lean Management is no longer limited to manufacturing, but is applied in other areas such as Innovation, Administration and Maintenance.

## Industrialized Tool Manufacturing

The Industrial Revolution affected almost all industrial sectors and increased both their productivity and technical capability[103]. Taylorism, Fordism and Lean Management delivered additional increases in efficiency by improving machine utilization, reducing inventory costs and minimizing throughput-time[104]. All of the paradigms presented in the previous Chapter, however, primarily targeted series production and thus have for many years seen only

---

[98] Cf. WOMACK, JONES AND ROOS (1990) The machine that changed the world: The story of lean production–Toyota's secret weapon in the global car wars that is revolutionizing world industry, pp. 1.

[99] Cf. KRAFCIK (1988) Triumph of the lean production system, pp. 1; WOMACK, JONES AND ROOS (1990) The machine that changed the world: The story of lean production–Toyota's secret weapon in the global car wars that is revolutionizing world industry, pp. 1.

[100]Kanban is a scheduling principle, which initiates the restocking of supplies based on their actual consumption.

[101]SMED is a method to reduce the changeover time necessary to switch from the production of one product to another product.

[102]VSM is a structured approach to recording and analyzing processes in order to identify waste.

[103]Cf. WIENDAHL, KUPRAT AND AHRENS Logistikgerechte Gestaltung von Produktionsstrukturen auf der Basis von Betriebskennlinien — Theorie und praktische Anwendung in der Metall- und Elektroindustrie, p. 226; AUST (1990) Ein Bewertungsverfahren für die Produktionsplanung bei auftragsorientierter Werkstattfertigung, p. 63.

[104]Cf. HERLITZ (1995) Lean Management als Wettbewerbsstrategie im deutschen Werkzeugmaschinenbau, p. 68.

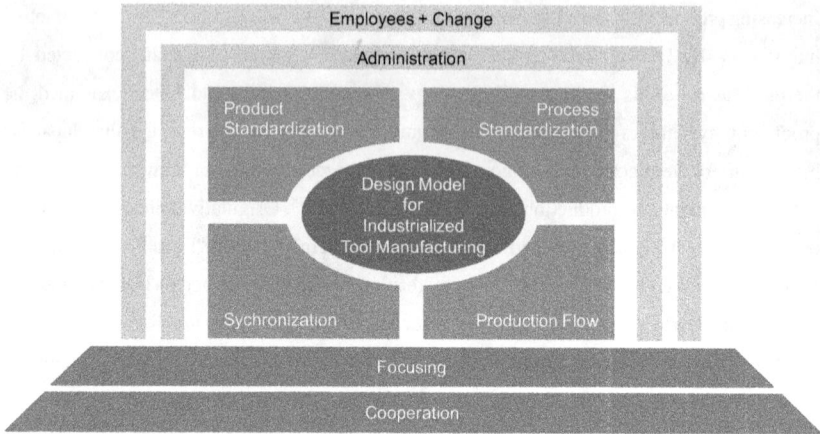

Figure 2.5: Design model for industrialized tool manufacturing

infrequent application in tool manufacturing. Application has been limited to the use of individual methods such as Poka-Yoke[105] or 5S[106].

The current focus on a "One-Piece-Flow" and high flexibility in series-production is, however, similar to requirements tool manufacturing sees itself confronted with[107]. In 2007 KLOTZBACH introduced the design model for the industrialization of tool manufacturing shown in Figure 2.5. It includes the core elements of industrialized production described above, but modifies them in accordance with requirements of the tool manufacturing sector.[108] The model consists of eight areas outlined below.

1. **Focusing**

   It is important for tool manufacturers to ensure their product spectrum is focused on their core competencies and products. Possibilities of achieving this can be either a reduction in the breadth of the offered product spectrum or a reduction of the depth of added value. These measures can help companies to remain competitive internationally, even in light of limited size and technological furnishings.

---

[105]Poka-Yoke is a method to avoid errors during production by designing products and processes to preclude these from occurring.
[106]5S is a set of rules to achieve and maintain order and cleanliness in production facilities
[107]Cf. KLOTZBACH (2007) Gestaltungsmodell für den industriellen Werkzeugbau, p. 2.
[108]Cf. Ibid., pp. 181.

2. **Cooperation**

   Cooperation with other tool manufacturers becomes necessary to ensure that a single manufacturer remains able to deliver complete tools to its customers after focusing his product spectrum. Since it is expedient to cooperate with partners who have core competencies other than those found in one's own company, the combination of cooperation and focusing holds the potential for considerable cost savings.

3. **Product Standardization**

   While tools are generally custom tailored to the demands of a customer and can thus be considered unique, the target modularization in conjunction with the standardization of the individual modules has proven to hold significant potential to improve efficiency along the entire value chain.[109] Furthermore by defining a set of standardized tool components, engineers can invest more time into the design of the non-standard components and thus produce a better product.

4. **Process Standardization**

   In tool manufacturing the production processes and the flow of materials are rather unstructured. This makes efficient production planning and systematic process improvement difficult[110]. Process standardization is not limited to the effects inherent in the standardization of components. Also the cross component standardization has proven to be possible and yields tangible advantages in practice[111].

5. **Production Flow**

   Based on a successful standardization of parts of the tool production processes, it is possible to identify common subsequences within these processes. This allows for the arrangement of production resources in the sequence necessary to perform certain aspects of tool development such as, e.g. soft- or hard-machining. This again reduces the necessary transport of materials on the factory floor and renders the production process easier to manage.

---

[109]Cf. BOOS (2008) Methodik zur Gestaltung und Bewertung von modularen Werkzeugen, pp. 138.
[110]Cf. MARCZINSKI (2008) Zeitschrift für den Betriebswissenschaftlichen Fabrikbetrieb, pp. 281.
[111]Cf. ZISKOVEN (2013) Methodik zur Gestaltung und Auftragseinplanung einer getakteten Fertigung im Werkzeugbau, p. 54.

6. **Production Synchronization**

   Synchronization of the tool manufacturing process was long deemed impossible due to the heterogeneity of the involved products. Work by ZWANZIG has shown, however, that this assumption is untrue. Instead of attempting to create a flow of materials for individual components, he focused on batches of components traveling through the production facility on similar paths. To reduce logistics on the factory floor these batches are moved from station to station simultaneously. While the principle has been shown to work very well and to yield significant gains in productivity the concept is still quite new in the tool manufacturing sector and thus is not yet in widespread use.[112]

7. **Administration**

   The successful implementation in the aforementioned areas requires modifications of the administrative processes as well. Sales processes can be restructured by referencing standardized components and thus simplifying the calculation. This also applies to controlling, where the added transparency allows for improved measurement of a tool production project's projected and actual cost. Furthermore, standardization allows for new opportunities regarding the management and development of external suppliers.[113]

8. **Workers and Change**

   The industrialization of tool manufacturing allows for the separation of simple and complex labor and thus for lowering of labor costs. This in turn allows higher paid workers to focus better on tasks requiring their more advanced skills. An example of this is the concentration of machine rigging efforts in one central area of the production facility. This would allow machine operators to focus better on their tasks. Although industrialization holds great potential it simultaneously represents a very significant shift in daily operations. Thus it is extremely important to invest resources into change management - especially in light of the traditional nature of tool manufacturing.

## Synchronized Tool Production

Synchronized production is a component of industrial tool manufacturing described in the previous Chapter. The specific synchronization concept for the tool manufacturing industry underlying this thesis is based on the work first pioneered by GRUSS and ZWANZIG in the

---

[112]Cf. ZWANZIG (2010) Taktung der Unikatfertigung am Beispiel des Werkzeugbaus, pp. 141.
[113]Cf. KLOTZBACH (2007) Gestaltungsmodell für den industriellen Werkzeugbau, pp. 178.

Mekropro[114] research project and later expanded upon by ZISKOVEN in InSynchroPro I[115116] and InSynchroPro II[117].

Arguments against synchronized tool manufacturing brought forth frequently in industrial practice are the wide variety of different products and non uniform capacity requirements along their respective production paths. On the surface the idea of a production system essentially based on the assembly line principle might seem to be a poor fit for the tool manufacturing industry. On the other hand, production process synchronization holds the potential to reduce throughput-time, improve adherence to delivery dates, reduce inventory costs and improve manufacturing system transparency. It was this potential that served as the motivation leading to the prior work referenced in the introduction of this Chapter.

When the principles of synchronization were applied to individualized production in general and tool manufacturing in specific, the basic idea of moving materials and semi-finished products through the manufacturing facility in a continuous, clock-driven flow was maintained. The unit of available production time, which is equal to the time between material movement operations is referred to as a *Tact*. Unlike conventional assembly line production however, the unit being moved in a constant rhythm is not an individual component, but rather a bundle or batch of components. The underlying idea was to create a bundle of components with relatively uniform capacity requirements from individual parts with non uniform capacity requirements. Thus given a suitable combination of components in a bundle, efficient use of the available production time within a Tact can be made.

The need to reduce the logistic costs associated with such a solution gave rise to the concept of the *Tactpallet* which is moved forward along a *Tactline* at the end of a Tact. This principle is shown in Figure 2.6 on the left[118]. The former refers to a physical container used to transport the aforementioned components from manufacturing step to manufacturing step. Once a Tactpallet commences its journey, its payload should no longer be modified. By bundling components and loading them onto a Tactpallet prior to commencing the manufacturing process multiple parts can be moved simultaneously with comparatively low

---

[114]IGF-Research Project 14424 N, Methodology for the Continuous Optimization and Assessment of Value-Adding Processes in the tool manufacturing industry
[115]Industry-funded venture of the WZL of the RWTH Aachen, the German industrial federation of massive forming and four tool manufacturers
[116]Cf. HINSEL Synchrone Fließfertigung: Auch in einem Werkzeugbau der Massivumformungen?, pp. 13.
[117]IGF-Research Project 16498 N, Industrialization and Synchronization of Production Processes in the Massive Forming Sector of the Tool Industry
[118]Based on ZISKOVEN (2013) Methodik zur Gestaltung und Auftragseinplanung einer getakteten Fertigung im Werkzeugbau, p. 89.

effort. On the shop-floor of tool manufacturers the use of wooden pallets for this purpose is widespread. A Tactline represents the sequence of destinations, to which a Tactpallet is transported when it passes through the manufacturing facility. As time and Tacts pass, the Tactpallet and thus the entire bundle are moved along the Tactline synchronously. The stations traversed along Tactline are referred to as *Tactstations*. In practice Tactstations are usually technological production steps such as turning, cutting, drilling or abrasive processes. Since manufacturing facilities often contain more than one physical machine of a given type, Tactstations again in turn consist of so-called *Tactsubstations*, which map directly to physical machines or workplaces. When a Tactpallet arrives at a Tactstation, the contained components are distributed among the available Tactsubstations for sequential processing. At the end of a Tact, the contents are gathered and loaded onto the Tactpallet for transport to the next Tactstation in the respective Tactline.[119] When a Tactpallet completes its journey along a Tactline, the contained finished or semi-finished products are unloaded. Subsequently they are either transferred to another Tactpallet, to an isolated processing step or to storage facilities. Upon having completed the final manufacturing operation those process steps which follow synchronized production, such as assembly or shipping are executed. Generally, production facilities contain not one, but multiple Tactlines which have multiple intersection points on physical machines or workplaces. Figure 2.7 shows a conventional tool production on the left and a synchronized production setup with multiple Tactlines on the right. Figure 2.6 on the right shows all of the introduced components constituting a synchronized production.

Driving the logistics of a production system based on a consistent Tact ensures a uniform flow of materials. While the bundling of multiple parts onto a Tactpallet automatically causes a certain idle time for individual components, it is considerably lower than the average idle time in a typical job shop setting. Data collected in the tool manufacturing industry shows that processing time typically accounts for between 1% and 5% of the total throughput-time[120]. Furthermore, strict application of the principle leads to a fully deterministic flow of materials. Provided that no unscheduled rework or machine failures occur, completion dates can be predicted with perfect accuracy, since the duration of a Tact and the number of Tactstations in a Tactline are known beforehand. The only logistical action necessary to sustain the flow of materials is the forwarding of Tactpallets between Tacts. Therefore the amount of logistics effort necessary on the shop-floor level is also considerably lower than in the case of traditional job shop manufacturing scenarios. An additional advantage on the shop-floor level is the high

---

[119]Cf. Ibid., p. 90.
[120]Based on a dataset containing approximately 300 elements from the past five years, which is stored in the tool manufacturing industry database at the WZL of the RWTH Aachen.

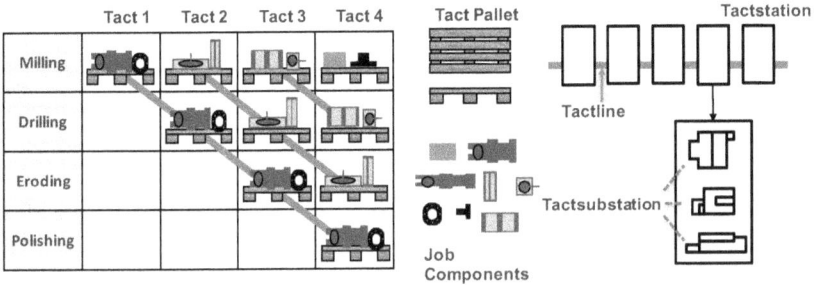

Figure 2.6: Components of the proposed synchronized production system

transparency and ease with which the flow of materials can be visualized and understood by production planners and workers alike.

While aforementioned advantages hold true in industrial practice, experience shows that only a certain percentage of jobs are suited for a synchronized production as described above. In the tool manufacturing industry there are two major reasons for this. Firstly there exists a fraction of jobs with required sequences of production steps so dissimilar to the other jobs constituting the bulk of the job spectrum, that these could not fill a suitable Tactline to capacity. Secondly, some jobs simultaneously have little or no lead time and high priority. Common examples in tool manufacturing are repair jobs which enable the commencement or continuation of a series production process (refer to Chapter 2.4). Due to the required processing with minimized wait-time, synchronized manufacturing with its fixed throughput-time is not necessarily a good fit.

The advantages of the synchronized production system described above do, however, come at a cost. Before being able to benefit from the manifold advantages during manufacturing,

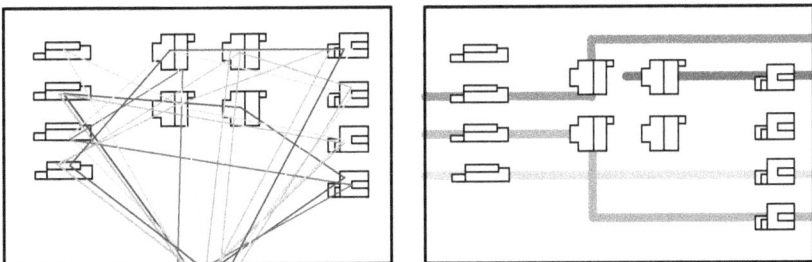

Figure 2.7: Principle of synchronized individual production

significant preparatory work must be invested. In the methodology developed by ZISKOVEN, this effort is structured as two major blocks. The first deals with the long term layout and production system setup while the second focuses on short term job scheduling and production planning[121]. While ZISKOVEN's work yielded impressive results in both blocks[122] industrial application revealed two areas necessitating further research:

1. In tool manufacturing, value chain production engineering identifies the sequence of manufacturing steps necessary to produce a tool component. In the context of synchronized production systems this sequence must however be translated into a sequence of Tactlines which is capable of performing the necessary production steps in the required order. When synchronized production setups mature, the number of contained Tactlines generally increases to provide a better coverage of the produced Job spectrum. Unfortunately, this renders the task of determining a suitable sequence of Tactlines for the production of a component complex and time consuming. Furthermore, this process effectively determines the minimal throughput time, and thus has a decisive influence on the performance of the production system. Since ZISKOVEN's work on short term production planning represents the state of the art and provides no algorithm capable of this translation tool manufacturing companies with synchronized production systems resort to time-intensive manual process.

2. The other area requiring research is the result of a limitation in ZISKOVEN's production scheduling algorithm. While the scheduling algorithm performed very well in industrial practice, it is limited to the scheduling of Jobs for a single Tactline. As mentioned in the previous paragraph, mature synchronized production setups, however, feature multiple Tactlines. Due to constrained resources these Tactlines often intersect on one or even multiple machines on the factory floor. This renders sequential executions of ZISKOVEN's algorithm for each Tactline impractical, since this could lead to the later Tactlines being "crowded out". Tactlines processed later would potentially not receive enough processing capacity to accommodate all jobs. It is for this reason that tool manufacturing companies resort to manual planning, once their Tactline Layout contains intersecting Tactlines. Since manual scheduling is exceedingly complex and difficult to grasp, it requires extensive amounts of time and often does not yield satisfactory results.

---

[121]Cf. Ibid., p. 93.
[122]Cf. Ibid., p. 193.

# Reflexion

Based on the research question posed in Chapter 1 this Chapter provided a description of the tool manufacturing as the area of application. Subsequently the topic of industrialization and its effects on tool manufacturing were discussed. Within the area of industrialized tool manufacturing the synchronization of production processes was elaborated upon. Examination of this in light of current industrial implementations revealed two specific issues not addressed by previous work. Thus the original research question posed in Chapter 1

*What activities are necessary to support the short-term planning of synchronized tool manufacturing, which takes due-dates into account, minimizes throughput-time and makes efficient use of machinery?*

can now be concretized yielding two new research questions

1. *What is a feasible design for an algorithm capable of **deriving a suitable sequence of Tactlines from a sequence of production steps generated by production engineering** while taking due-dates, the minimization of throughput-time and efficient use of machinery into account?*

2. *What is a feasible design for an algorithm capable of **scheduling jobs in a synchronized production system with multiple, intersecting and concatenated Tactlines** while taking due-dates, the minimization of throughput-time and efficient use of machinery into account?*

These questions are reflected in the three major components of the methodology *Tactsequence Mapping, Production Planning*s presented in this thesis and described briefly in the following sections.

# Chapter 3

# Concept Outline

Building upon the two questions guiding the research in this thesis, this presents an outline of the two resulting optimization problems referred to as Tactsequence Mapping and Production Scheduling. Furthermore this Chapter introduces the notation used to describe the algorithms presented in later Chapters of this thesis.

## Tactsequence Mapping

This Chapter will present an outline for Tactsequence Mapping and thus propose a solution to the first research question raised at the end of Chapter 2.4.

Since the focus of this thesis is short term production planning, necessary preceding steps are assumed to be complete. Thus the analysis of the process spectrum, the determination of Tactlines, Tactstations and Tactsubstations as well as the specification of the Tact-duration i.e. Tact-Rhythm have been completed successfully. The result is a fully defined synchronized production system. With all of this in place, the first essential task prior to the commencement of regular production planning activities is the assignment of a suitable sequence of Tactstations. This is a necessary prerequisite if a given job is to be manufactured within the confines of the predefined synchronized production system. The necessary activities are combined in *Tactsequence Mapping*, the first module of the methodology presented in this thesis.

Upon completing the construction step of tool development (refer to Chapter 2.4), all components constituting a tool are known, their designs are complete and respective due-dates have been derived. Thus for the purposes of manufacturing a tool, an order is a set of interdependent jobs for the individual tool components. To be able to include these in the production scheduling process, the series of production steps necessary for each job must be mapped to a sequence of *Tactstation-Requirement*, hence referred to as *Tactsequence*.

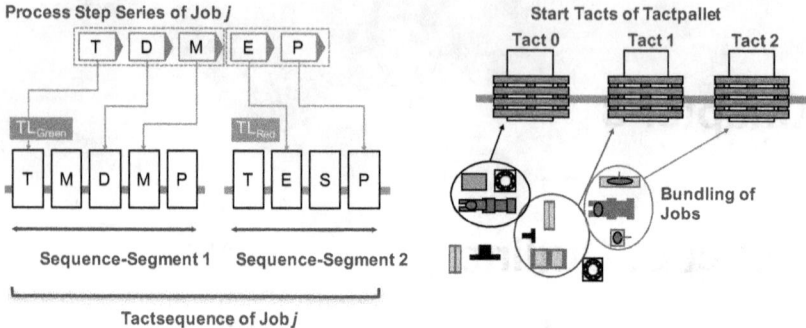

Figure 3.1: Tactsequence Mapping and Production Scheduling working principles

A Tactsequence is essentially a sequential arrangement of Tactstations which a job must pass through in the course of its manufacturing process. Within the full Tactsequence, the contained Tactstations are grouped into **Sequence-Segment** which represent contiguous sections of the sequence within which a job must move from Tactstation to Tactstation without pause or interruption. A Sequence-Segment can consist of either one or multiple Tactstations. In essence Sequence-Segments are the manifestation of passing through an entire Tactline (multiple Tactstations) or receiving isolated processing on a separate machine (single Tactstation). Figure 3.1 on the left shows an example in which a job first passes through *Tactline Green* and finally completes manufacturing by passing through *Tactline Red*. The example illustrates both the activity of Tactsequence Mapping and the underlying concepts such as Job, Tactsequence, Sequence-Segment, Tactline and isolated processing.

Even for simple production system configurations there is often no functional relationship between jobs and Tactsequences. Thus jobs can theoretically be mapped to multiple different Tactsequences. These can however differ in many respects such as throughput-time, associated logistics operations and number of isolated processing steps. Thus these are considered to be more or less "good" in light of the exogenous objectives and restrictions, e.g. "minimization of inventory", "minimization of throughput-time", "available logistics infrastructure" and "available lead-time". Thus choosing the mapping which is the best fit for the production environment is a requirement from industrial practice which was taken into account when developing the Tactsequence Mapping methodology presented in this thesis.

# Production Scheduling

After having provided an outline for an answer to the first research question raised at the end of Chapter 2.4, this Chapter will attend to the second research question and thus present an outline for Production Scheduling.

Upon completion of Tactsequence Mapping, jobs have been assigned a Tactsequence which allows for synchronized sequential completion of all necessary manufacturing steps. The synchronization concept outlined in Chapter 2.3.3 is based upon the idea, that given an optimized bundling of components on Tactpallets, efficient use of the available production time within a Tact can be made. Simultaneously this allows for a consistent flow of materials through the production facilities. *Production Scheduling*, the second module of the methodology presented in this thesis, serves to assemble these bundles and assign them to certain Tactpallets or isolated processing steps during a specific Tact. Figure 3.1 on the right illustrates the principle. This assignment must be carried out taking the adherence to job due-dates into account and simultaneously maximizing machine utilization while minimizing throughput-time and inventory[123].

Due to the inherent conflict of objectives, scheduling while simultaneously optimizing all of the aforementioned parameters is impossible[124]. Inherent in the nature of the synchronization concept introduced above, the influence on throughput-time during production scheduling is limited to adjusting the time jobs spend in storage between Sequence-Segments. Since the remaining throughput-time of a job is equal to the number of Tactstations in its Tactsequence multiplied by the duration of a single Tact, greater influence on throughput-time can be exerted during Tactsequence-Mapping or during the setup of the production system.

This leaves two remaining parameters for optimization: **adherence to due-dates** and **maximization of machine utilization**. In line with prior work, the methodology and algorithms presented in this thesis will give adherence to due-dates priority over machine utilization. The primary reasons for this choice lie in the increased industry focus on *just-in-time* and *just-in-sequence* supply of parts[125].

---

[123]Cf. (2010)Operative Exzellenz im Werkzeug- und Formenbau, pp. 16.
[124]Cf. VAN DYKE PARUNAK (1991) Journal of Manufacturing Systems, p. 243.
[125]Cf. HÜTTMEIR et al. (2009) International Journal of Production Economics, p. 501; BRAKEMEIER AND JÄGER (2004) Schlanke Produktion, p. 86; ZISKOVEN (2013) Methodik zur Gestaltung und Auftragseinplanung einer getakteten Fertigung im Werkzeugbau, p. 103.

Prior work by ZISKOVEN[126] provides an excellent foundation upon which to build when solving this problem. Yet the increased complexity of the synchronization concept described above renders direct application of the algorithms contained therein impossible. Since a job can now pass through a series of Tactlines and isolated processing steps along its Tactsequence, the production scheduling must support the sequential chaining of Sequence-Segments (refer to Chapter 5.2) each of which represent Tactlines with multiple potential intersection points (refer to Chapter 5.2). Iterative approaches, in which individual segments are scheduled upon completion of preceding segments, represent a simpler scheduling problem to solve[127]. As a matter of principle these can however not provide assured adherence to due-dates. To ensure timely job completion under these conditions[128], scheduling must thus be performed simultaneously for all Sequence-Segments along a Tactsequence.

Furthermore the repeated application of the scheduling algorithm to multiple intersecting Tactlines in a production facility as suggested by ZISKOVEN[129] is impractical in industrial practice. Since no a-priori knowledge regarding a suitable distribution of available machine capacity between the involved Tactlines exists, scheduling would consist of a series of locally optimized passes where the number of necessary iterations is an exponential function of the number of Tactline intersections in a Sequence. Clearly such an approach is not feasible and thus warrants integrated processing of jobs across all Tactlines.

## Notation

This Chapter gives an overview of the notation used to describe the manipulation of relations and tuples during algorithm execution in Chapter 5. Since both *relational Algebra* and *Relational Calculus* are equally expressive languages[130] these are henceforth used interchangeably depending on their situational suitability and conciseness. Beginning with a short overview of the Relational Model a detailed description of the two aforementioned languages is then provided.

---

[126]Cf. Ibid..
[127]Cf. Ibid., p. 101.
[128]precluding unforeseeable disturbances to the production process
[129]Cf. Ibid., p. 184.
[130]Cf. CODD Relational Completenes of Data Base Sublanguages, p. 1.

# Relational Model

The **Relational Model** ($RM$) was introduced by CODD in 1969 as means of describing and structuring data[131]. Due to its foundation in mathematical theory, disciplined data organization, simple and familiar view of data and the means of establishing logical relationships the RM has since established itself as a superior logical data model[132]. An important concept underlying the relational model is that of a well-defined collection of objects henceforth referred to as a *set*[133].

Thus with $n$ sets $S_1, S_2, ... S_n$ the Cartesian product of these sets is defined as

$$S_1 \times S_2 \times ... \times S_n = \{(s_1, s_2, ..., s_n) | s_1 \in S_1, s_2 \in S_2, ..., s_n \in S_n\}. \qquad (3.1)$$

Elements in the resulting set are designated as *tuples*. According to the mathematical definition, a subset of the relation between two sets is defined as a subset of their Cartesian product[134]. A more practical way of perceiving relations consisting of tuples is as tables of data containing rows. Unlike conventional tables however the tuples in a relation have no specific order[135]. The "columns" in a relation, derived from the aforementioned underlying sets, are referred to as *attributes*.

A Relation $R$ with the exemplary attributes Name, Description and Weight is denoted as

$$R = \{Name, Description, Weight\} \qquad (3.2)$$

Since the intuitive view desired by the user is often prone to so-called anomalies it is thus less than ideal from the perspective of information theory. The problem can be alleviated by decomposing relations with anomalies to well-structured relations with minimal redundancies[136]. CODD introduced the process referred to as *normalization*, which is now used widely when designing and structuring data stores[137].

---

[131]Cf. CODD (2009) SIGMOD Rec, pp. 17.
[132]Cf. PONNIAH (2007) Data modeling fundamentals: A practical guide for IT professionals / Paulraj Ponniah, pp. 231.
[133]Cf. DEVLIN (2003) Sets, functions, and logic, p. 57.
[134]Cf. FRAENKEL, BAR-HILLEL AND LÉVY (1973) Foundations of set theory, pp. 41.
[135]Cf. FONG (2015) Information systems reengineering, integration and normalization, p. 44.
[136]Cf. SUMATHI AND ESAKKIRAJAN (2007) Fundamentals of relational database management systems, p. 296.
[137]Cf. CODD (1970) Communications of the ACM, pp. 377; CODD Further normalization of the data base relational model, pp. 245; CODD (1971) Normalized data base structure, pp. 1.

# Relational Algebra

*Relational Algebra* ($RA$) introduced by CODD in 1972 is a procedural query language[138] consisting of a number of mathematical operations that serve to compare, modify and combine tuples within different relations. Thus Relational Algebra provides the tools necessary to filter and compose the data contained in normalized relations in accordance with a user's wishes.[139] When it was first introduced in 1972 RA included seven primary operators, which are covered in the first part of this Chapter. Over the years the number of operators has however increased. The expansions are described in the second part of this Chapter. Finally some more complex queries used frequently in the context of this thesis are included in the third and final part of this Chapter.

## Basic

In the interest of succinctness and since the basic operators were all defined in CODD's seminal paper[140], no additional references to this paper will be included in this Chapter.

The $\times$ operator is the *Cartesian Product* commonly also referred to as the *Cross Product*. The version included in RA functions on relations in a similar way to its operation on sets (refer to Chapter 3.3.1). Given two relations $A$ and $B$ the cross product

$$R = A \times B \qquad (3.3)$$

pairs each tuple in $A$ with each tuple in $B$ and places the result in a new relation $R$. The new relation contains the sum of all attributes in $A$ and $B$. Should $A$ and $B$ contain attributes with identical names, these attributes are added to $R$ with prefixes denoting their origin relation (e.g. $A.Attribute$ and $B.Attribute$).

In addition to the Cartesian Product three additional binary operators commonly used on sets are included in the set of basic RA operators. Given two relations $A$ and $B$ the three operators

$$R = A \cup B \quad : \quad \text{Union} \qquad (3.4)$$

$$R = A \cap B \quad : \quad \text{Intersection} \qquad (3.5)$$

$$R = A - B \quad : \quad \text{Difference} \qquad (3.6)$$

---

[138]Procedural languages express a sequence of operations, which are to be applied in a certain specified order.
[139]Cf. CODD Relational Completenes of Data Base Sublanguages, p. 1.
[140]Cf. Ibid., pp. 1.

*Union, Intersection* and *Difference* function as expected. Thus they serve to combine two relations, yield only identical elements in both relations and yield only those elements in A, which are not in B, respectively. The only difference when compared to their mathematical counterparts lies in the fact that these apply only to compatible normal relations.

The *Project Operator* $\pi$ is an operator which serves to pick only specific attributes from a relation. Given a relation $A$ and a list of relevant attributes $L$, the project operator

$$R = \pi_L(A) \cup B \tag{3.7}$$

yields a new relation with the attributes contained in $L$. Thus the effect of applying the Project Operator is comparable to selecting only a certain subset of relevant columns in a table. An additional usage of the Project Operator can be to calculate new attributes from existing attributes by using simple arithmetic expressions in $L$

When only a certain subset of the tuples contained in a relation is of interest, the *Restriction Operator* $\sigma$ serves to filter data accordingly. Given a relation $A$ and a restriction $S$, the Restriction Operator

$$R = \sigma_S(A) \tag{3.8}$$

yields a new relation containing only those relations in $A$ which conform to the restriction $S$. A restriction in this case is an expression containing $=, \neq, <, \leq, >$ or $\geq$. Similar to the Projection operator's effect, the Restriction Operator selects only a certain subset of rows in a table.

When working with normalized data, data is generally distributed across several relations. The *Join* operator $\bowtie$ which exists in two variations serves to undo separation by combining two relations. Given two relations $A$ and $B$

$$R = A \bowtie B \tag{3.9}$$

$$R = A \underset{\theta}{\bowtie} B \tag{3.10}$$

$\bowtie$ yields a new relation with a combination of the attributes in $A$ and $B$. The tuples in the result are the elements of the Cartesian Product of $A$ and $B$ where those attributes contained both in $A$ and $B$ are equal (Equation (3.9)) or where the expression $\theta$ is satisfied (Equation (3.10)). Thus the Join operator is equivalent to $\sigma_E A \times B$ where $E$ expresses the equality requirement for all attributes both in $A$ and $B$ in the first case or $E$ is equal to $\theta$ in the second. When applied without the $\theta$ expression this operator is also often referred to as a *Natural Join*.

The *Division* operator $\div$ filters a relation by returning only those tuples from the first operand, which are also contained in the second operand. Thus given two relations $A$ and $B$

$$R = A \div B \tag{3.11}$$

yields a new relation $R$ which only contains those tuples from $A$, where all tuples in $B$ are fully contained in $B$. For this operator to function, it is important to note, that the attributes in $B$ must be a subset of the attributes in $A$.

## Extended

The following operators were not included in the original definition but were added over the years and are now in widespread use and documented in a variety of books [141]. This Chapter introduces four operators, which are used frequently in the later Chapters of this thesis.

The *Antijoin* operator $\triangleright$ is used to isolate those element in a relation, where the other relation contains no corresponding elements. Given two relations $A$ and $B$

$$R = A \underset{\theta}{\triangleright} B \tag{3.12}$$

yields a new relation $R$ which contains those elements from relation $A$, where no corresponding elements exist in relation $B$ when compared with the comparison expression $\theta$.

The *Rename* operator $\rho$ is used to rename the attributes of a relation. Generally the argument given in the subscript is used to denote the new attribute names to be used after the rename operation in comma-separated form (Equation (3.13)). While this syntax is used on occasion in the context of this thesis, the operator is mostly used with a slightly different syntax in the interest of succinctness (Equation (3.14)). Given a relation $A$

$$R = \rho_{N_1, N_2, \dots, N_k}(A) \tag{3.13}$$

$$R = \rho_{O/N}(A) \tag{3.14}$$

yields a new relation $R$, where a single attribute $O$ is renamed to $N$. While this notation limits the operation to the renaming of only one attribute, this form allows for significantly shorter and thus more comprehensible notation in later Chapters.

---

[141]Cf. ABITEBOUL, HULL AND VIANU (1995) Foundations of databases, pp. 91; KEDAR (2009) Database Management System, pp. 3-1; SUMATHI AND ESAKKIRAJAN (2007) Fundamentals of relational database management systems, pp. 76.

When a relation is to be purged of duplicate tuples, the *Deduplication* operator $\delta$ can be applied to a single relation $A$

$$R = \delta(A) \tag{3.15}$$

yielding a new relation $R$ containing only unique tuples.

The *Aggregation* operator serves to apply one of several aggregation functions to the tuples in a relation and adds an additional attribute containing the result. Thus given a single relation $A$

$$R = \gamma_{O_1,...,O_n,\vartheta} A \tag{3.16}$$

first applies the aggregation function defined as $\vartheta$ to the tuples in the relation where the attribute values for attributes in $(O_1, ..., O_n)$ are identical. In a next step the attributes listed in $(O_1, ..., O_n)$ are projected, the resulting relation is then deduplicated and finally the results gathered during the application of $\vartheta$ are added to the tuples in the resulting relation as an additional attribute. Literature lists a variety of aggregation functions such as

$$count(), sum(O), avg(O), stdd(O), max(0), min(O) \text{ or } order(O). \tag{3.17}$$

While *count* calculates the number of tuples in the relation it is applied to *sum, avg,stdd, max* and *min* calculate the sum, average and standard deviation, maximum and minimum of the tuples' attribute values indicated by the argument $O$. The order function finally calculates the relative order of tuples if they were sorted in ascending order by the attribute indicated by the argument $O$.

## Complementary

This Chapter contains helpful complementary functions which are used throughout Chapter 5. While these functions are merely shorthand for more elaborate expressions, their use significantly improves readability.

The *first* function returns an the an arbitrary element from a relation. Given a non-empty relation $A$ with a set of attributes $B$

$$first(A) = \pi_B(\sigma_{order(B)=1}(\gamma_{B,order(B)}A)). \tag{3.18}$$

The *corresponds* function $\psi$ checks if two relations contain corresponding elements in a specified order. Given two relations $A$ and $B$ with $O_1$ and $O_2$ as the sets of attributes to order by and with $C_1$ and $C_2$ as sets of comparison attributes

$$\psi_{C_1,C_2}^{O_1,O_2}(A,B) = \gamma_{O_1,order(O_1)}(A) \underset{order(O_1)\cup C_1=order(O_1)\cup C_2}{\triangleright} \gamma_{O_2,order(O_2)}(B) \tag{3.19}$$

returns those tuples in $R_1$ where $R_2$ does not contain at least one tuple with matching $(C_1), (C_2)$ per ordered-group created by aggregating on the attributes $O_1$ and $O_2$.

# Relational Calculus

Similar to RA defined above, **Relational Calculus** ($RC$) is a mathematical notation, with which relations can be queried and manipulated. The purpose of RC at its inception was however to represent a formal basis for the definition of query languages. Unlike RA Relational Calculus falls into the category of declarative[142] languages, where the focus is not to "state *how* to perform a certain task" but rather on "describing *what* task to perform". While in general there are different types of RC, the usage in this thesis is limited to so-called Tuple Relational Calculus (TRC) which will be introduced in condensed form in the first part of this Chapter.[143] The second part of this Chapter contains several queries used throughout the thesis.

## Tuple Relational Calculus

Introduced in conjunction with the relational model in 1972 by CODD, Tuple Relational Calculus consists of expressions which take the following form:

$$\{\text{<tuple variable(s)>} \mid \text{<logical formula>}\}. \tag{3.20}$$

An example of a query "find all tuples in a relation named $cars$ which feature a power-attribute $power$ greater than or equal to 218" could then be formulated as:

$$\{C \mid (C \in cars) \land (C.power \geq 218)\}. \tag{3.21}$$

Tuple variables declared before the "pipe"[144] are hence referred to as *free*. In general free variables apply to tuples in all relations where the logical formula specified after the "pipe" is satisfied. For this reason the first part of the expression in Equation (3.21) narrows the scope

---

[142]Formally the "term *declarative programming* stands for the combination of functional (or applicative) and relational (or logic) programming" (cf. PADAWITZ (1992) Deduction and declarative programming, p. 1).
[143]Cf. SUMATHI AND ESAKKIRAJAN (2007) Fundamentals of relational database management systems, p. 90.
[144]A vertical line is also referred to as a "pipe" character.

of the all tuples to test for the ensuing condition to those contained in the *cars* relation. The set of all tuples to which the logical formula applies then represents the result of a TRC query.

To allow for querying effectively across multiple relations, the logical formula can also contain so-called *bound* variables, which are connected to one of two following types of quantifier

∃: the *exists* quantifier is *true* if at least one tuple in the query scope exists, which would make the logical formula true

∀: the *for all* quantifier is *true* if all tuples in the query scope would make the logical formula true.

By using these quantifiers, it becomes possible to phrase more complex questions such as "find all tuples in a relation named *cars* where at least one tuple in the *owner* relation, linked to *cars* by the attributes $CarID$ and $CarID$, exists, which has an age greater than 32":

$$\{C \mid C \in cars \wedge \exists O((O \in owner) \wedge (C.CarID = O.CarID) \wedge (O.age > 32))\}. \quad (3.22)$$

TRC thus provides an alternative language when phrasing queries, which in some cases is more (e.g. Equations (3.23) and (3.24)) and in some cases less (e.g. Equation (3.9)) succinct than Relational Algebra. Since both RA and TRC operate on and modify relations, these two languages can and will be used interchangeably in this thesis.

## Complementary

As was the case for the corresponding Chapter when dealing with Relational Algebra, this Chapter contains helpful complementary functions which are used throughout Chapter 5.

The *max* and *min* functions retrieve the tuple, where an attribute value either has the maximum or minimum value in the parent relation. Given a relation $A$ and a single attribute $S$

$$min_S(A) = \{t \mid t \in R \wedge (\forall k)((k \in R) \wedge (t.A \leq k.A))\} \quad (3.23)$$

$$max_S(A) = \{t \mid t \in R \wedge (\forall k)((k \in R) \wedge (t.A \geq k.A))\} \quad (3.24)$$

*min* and *max* retrieve the entire tuple with the maximum or minimum attribute values.

The *Sequence* function generates a set of integers from 1 to $n$. Thus given an integer $n$ and a relation *integers* containing all integers from $-\infty$ to $\infty$:

$$sequence(n) = \{y \mid (y \in integers) \wedge (y \geq 0) \wedge (y < n)\}. \quad (3.25)$$

# Chapter 4

# Theoretical Fundamentals

This Chapter presents necessary theoretical fundamentals by first providing a background in the area of combinatorial optimization. Subsequently the sequence alignment and knapsack problems from the areas of computer science and operations research respectively are presented. These will be used as basic tools to construct the algorithms in Chapter 5.

## Combinatorial Optimization

The objective of mathematical programming, often also referred to as mathematical optimization, is the formulation and solution of constrained optimization problems[145]. In other words mathematical optimization entails both the definition of a mathematical target function subject to a set of constraints and the subsequent determination of the optimal solution subject to the aforementioned constraints. Combinatorial Optimization performs the same task, but does so taking the complication of indivisible activities, i.e. variables which can assume integer values only, into account[146]. While this additional limitation serves to create a finite set of possible solutions, this set is typically huge or more precisely grows exponentially in the number input variables[147]. Thus the sheer size of the solution space raises the question of how to solve this class of discrete optimization problems. This Chapter will provide a brief overview of the complexity of combinatorial optimization followed by approaches to both, i.e. solve problems precisely and approximate their solutions.

## Complexity

The exponential growth of the solution space for combinatorial problems renders evaluation of all possible solutions to select the best one an infeasible option. This raises the question

---

[145]Cf. SNYMAN (2005) Practical mathematical optimization: An introduction to basic optimization theory and classical and new gradient-based algorithms / by Jan A. Snyman, p. 1.
[146]Cf. (1980)Combinatorial Optimization, p. V.
[147]Cf. SCHRIJVER (2003) Combinatorial Optimization: Polyhedra and Efficiency, p. 1.

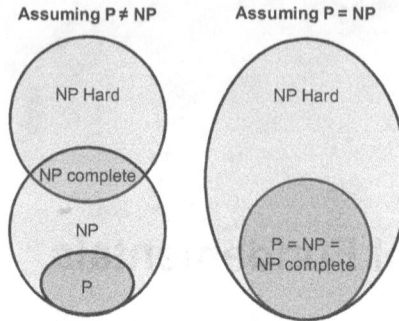

Figure 4.1: Overview of problem complexities

of practical computability, which was first posed by GÖDEL in his letter to VON NEUMANN in 1956. In this letter GÖDEL wonders if combinatorial problem could be solved in linear quadratic or even linear time.[148] In the 1960s John Edmonds introduced several algorithms, which were able to find the optimal solution to a combinatorial problem in polynomial time. Based thereupon he proposed, that all such algorithms be referred to as efficient.[149] In 1971 COOK explicitly phrased the question of whether such an efficient algorithm could be found to every conceivable combinatorial problem[150].

Before continuing with the discussion of computational complexity a few definitions are now in order. Let NP (*nondeterministic polynomial time*) be defined as the set of all problems for which the correctness of a potential solution can be efficiently checked. Basically this definition applies to all of the combinatorial problems examined in this thesis. Furthermore let P (*polynomial time*) be the subset of NP for which efficient, polynomial time solution algorithms exist. Finally let NP-hard be the set of all problems which are typical of the hardest problems in NP[151].

Using these definitions COOK's question can be rephrased to whether $P = NP$ holds[152]. The answer *no* ($P \neq NP$), while intuitive and supported by both philosophical and empirical considerations, however remains unproven to date[153]. Assuming for the moment, that this

---

[148]Cf. HARTMANIS (1989) Bulletin of the European Association for Theoretical Computer Science, pp. 101.
[149]Cf. SCHRIJVER (2003) Combinatorial Optimization: Polyhedra and Efficiency, p. 1.
[150]Cf. STEPHEN A. COOK (1971) The complexity of theorem-proving procedures, pp. 151.
[151]*Typical of the hardest problems in* $NP$ in this case means, that given a problem $X$, the class of problems $NP$ and a suitable class of reductions $C$. If $Y \leq_R X$ for all $Y \in C$ then a problem is considered to be $NP$-hard (cf. VAN LEEUWEN (1994) Handbook of Theoretical Computer Science: Vol. A: Algorithms and Complexity pp. 80). Put more simply this means a problem, which all problems in NP can be reduced to given a suitable reduction in $R$ are considered $NP$-hard
[152]Cf. GOLDREICH (2010) P, NP, and NP-completeness: The basics of computational complexity, p. 48.
[153]Cf. Ibid., p. 69.

answer is correct, this would entail there being another subset in $NP$ to account for the difference of $P$ and $NP$ ($NP - P$). This set of problems would then simultaneously be representative of the most difficult problems in $NP$ and yet be contained in $NP$. This set of problems is commonly referred to as $NP$-complete. If COOK's question could now be answered with *yes* this would mean that $NP = P = NP$-complete. Until that is the case, the conclusion is that $NP$ contains problems in $NP$-complete for which there probably exists no efficient algorithm to find the optimal solution.

An additional classification within the set $NP$-complete is the differentiation between *strongly* and *weakly* $NP$-complete problems. The latter are not as intractable as the former. More specifically weakly $NP$-complete can be solved exactly using pseudo-polynomial time algorithm if the input length and the magnitude of the largest number in the problem instance are bounded by a polynomial[154]. Thus these exhibit exponential behavior only when the problem instance grows exponentially in size[155]. In his seminal paper GAREY goes to show that for strongly $NP$-complete problem variants, a pseudo-polynomial algorithm cannot exist, except if $NP = P$.[156] A common example of a weakly $NP$-complete is the Binary Knapsack Problem discussed in Chapter 4.3.1. Interestingly several of its derivatives discussed in Chapter 4.3.2 are however strongly $NP$-complete.

## Finding Optimal Solutions

Depending on the encountered problem and the size of the relevant instances there are several viable approaches to solving combinatorial optimization problems in practice. Efficient algorithms to solve combinatorial problems in $P$ are generally problem specific. As these do not apply to the problems encountered in later Chapters of this thesis, a discussion of these will be omitted[157]. Thus the objective of this Chapter is to present strategies and practical approaches useful when solving $NP$-complete problems. Due to the exponential growth of the available solution space (with increasing problem size), explicit checking of all possible solutions with a brute-force approach is infeasible for all but the smallest problem instances. While the approaches discussed below cannot be used to solve $NP$-complete combinatorial problems in polynomial time on their own, the underlying ideas often represent the basis for

---

[154]Cf. GAREY AND JOHNSON (1978) Journal of the ACM, pp. 500.

[155]Cf. GAREY AND JOHNSON (1979) Computers and intractability: A guide to the theory of NP-completeness / Michael R. Garey, David S. Johnson, p. 91.

[156]Cf. GAREY AND JOHNSON (1978) Journal of the ACM, pp. 499.

[157]An example of problems simultaneously in $NP$ and $P$ is the *Continuous Knapsack Problem*, which can be solved in linear time using a weighted median search. For more information on the subject please refer to KORTE AND VYGEN (2008) Combinatorial optimization: Theory and algorithms.

problem specific algorithms with reasonable real-world performance (refer to Chapter 4.2 and Chapter 4.3).

## Branch-and-Bound

Branch-and-Bound represents a framework for finding the optimal solution for a discrete linear optimization problem. It was presented by LAND AND DOIG in 1960[158]. While it does not specify a recipe to solving a specific combinatorial optimization problem, it represents a generic tool, which can be adapted to solve a wide range of problems with greater efficiency than attempting a brute force solution would allow. Common applications include assignment-, sorting-, grouping- and choice problem. The basic idea - similar to the divide-and-conquer approach[159] - is to solve a very hard problem by breaking it down into a hierarchical set of subproblems and by systematically solving a subset of these recursively to arrive at the optimal solution for the superordinate problem[160]. Although this technique cannot transform a $NP$-hard problem into a problem solvable in polynomial time, depending upon the characterization it can dramatically reduce the necessary computational time to calculate the optimal result. On an abstract level, the approach involves three steps: branch, bound and prune, which will now be described in greater detail.

A problem $A_0$ is first broken down into multiple subproblems $A_i (i = 1, ... n)$ (branch), by partitioning the solution space of two or more disjunct sets using a branching function $B(A)$.

With $n \geq 2$

$$B(A_0) = \bigcup_{i=1}^{k} B(A_i) \tag{4.1}$$

while

$$B(A_i) \cap B(A_j) = \emptyset \tag{4.2}$$

for $i \neq j$

The first branching operation created child-nodes for the comprising elements. When this process is repeated, the next node for branching must be chosen. In this context a variety of different branching strategies exist, which range from simple depth- or breadth-first search to choices, which choose a branch target based on an appropriate heuristic. When performed

[158]Cf. LAND AND DOIG (1960) Branch-and-Bound Applications in Combinatorial Data Analysis, p. 497.
[159]Cf. CORMEN (2009) Introduction To Algorithms, p. 65.
[160]Cf. WEINBERG (1968) Einführung in die Methode Branch and Bound: Unterlagen für einen Kurs des Instituts für Operations Research der ETH, Zürich, p. 5.

exhaustively, the approach described thus far is a systematic form of full enumeration. There are however also strategies, in which the algorithm terminates prior to calculating the true optimal solution and thus delivers only an approximation thereof.

To reduce the number of necessary branching steps, the objective now is to show that one or multiple of these partitions cannot possibly contain the optimal solution to the problem. Thus the need for full enumeration of said partition is eliminated and the remaining effort necessary to solve the problem is reduced. One possible way of making this determination is by comparing the best possible solution of a subproblem $A_i$ to a known best valid solution (bound).

This best known valid solution, hence denoted as $L$, is initially set to $L = \infty$ or a solution calculated experimentally[161]. The best possible solution in this case can either be computed exactly trivially or otherwise estimated using a heuristic. Should trivial calculation yield a solution $L_{A_i}$, which is in compliance with all problem constraints, it is compared with $L$ and replaces it, should $L_{A_i}$ be a better solution. Should however, estimation of a subproblem's best possible solution be necessary, solution feasibility, i.e. conformance to the to all constrains, is nonessential. The only necessary condition is that for all possible problems $A_i$ a solution must be computed, which is better or equal to the true optimal solution of the problem. Thus to ease computability, the problem's constraints can therefore be relaxed. In the case of discrete linear optimization problems relaxation of the integrality constraint (LP-relaxation) allows for solution calculation in polynomial time. Since the optimal solution $SB_{A_i}$ computed with relaxed restrictions cannot be worse than that of the non-relaxed problem, it represents the upper bound of solution goodness for the respective subproblem.

After calculation of bounds and or feasible results, the entire set of non-evaluated subproblems can be usually purged from superfluous elements. By removing those subproblems, which have a best possible solution inferior to the best known valid solution can safely be pruned from the tree of problems to be evaluated to reduce the remaining number of calculations without sacrificing the solution optimality.

## Dynamic Programming

First introduced in the 1940s by the American mathematician RICHARD BELLMAN, the Dynamic Programming paradigm is a method designed to solve multi-stage decision

---

[161]Cf. PARKER AND RARDIN (1988) Discrete Optimization, p. 159.

processes[162]. In practice Dynamic Programming has proven a valuable tool with which a wide array of optimization problems can be solved with relative ease. By intelligently breaking down a problem into simpler subproblems[163] and evaluating all possible solutions it pursues a strategy similar to that at the heart of Branch-and-Bound (refer to Chapter 4.1.2.1).

Problems solvable by Dynamic Programming must exhibit an optimal substructure, i.e. the optimal solution must in turn consist of optimal solutions to contained subproblems[164]. Unlike the rather brutish Branch-and-Bound a set of constituting subproblems is first identified for a problem. Afterwards the simplest unsolved problems are selected, solved and the results are stored for later usage. Building thereupon solutions for more complex problems are then calculated iteratively[165].

In contrast to the divide-and-conquer approach[166] the residual subproblems are relatively large. Thus the tree structure resulting from Dynamic Programming problem decomposition often has a polynomial depth in contrast to the usually logarithmic depth of divide-and-conquer trees. The exponential growth in the number of nodes with increasing tree depth results in a massive increase in the number of nodes or subproblems to be calculated. Since many of these subproblems are however identical, saving their results for later usage greatly increased time and space efficiency.[167]

Figure 4.2 shows an example in which the recursive and dynamic programming approaches when calculating the Fibonacci numbers are contrasted[168]. The sequence is formally defined as

$$f(n) = f(n-1) + f(n-2) \qquad (4.3)$$

with $f(0) = 0$ and $f(1) = 1$ for $n > 1$.

The recursive evaluation of this formula is shown in tree-from Figure 4.2 on the left. Closer inspection reveals that this mode of calculation requires a high level of operations and has an execution time which is exponential in $n$. The right side shows the corresponding Dynamic

[162]Cf. BELLMAN (1972) Dynamic Programming, p. vii.
[163]Cf. LUUS (2000) Iterative dynamic programming, p. 12.
[164]Cf. CORMEN (2009) Introduction To Algorithms, p. 425.
[165]Cf. DASGUPTA, PAPADIMITRIOU AND VAZIRANI (2008) Algorithms, p. 170.
[166]Cf. CORMEN (2009) Introduction To Algorithms, p. 65.
[167]Cf. DASGUPTA, PAPADIMITRIOU AND VAZIRANI (2008) Algorithms, pp. 173.
[168]In 1202 Italian mathematician LEONARDO FIBONACCI USED A SEQUENCE TO MODEL THE GROWTH OF RABBIT POPULATIONS IN HIS BOOK Liber abbaci.

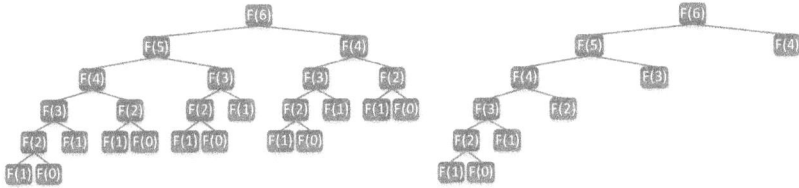

Figure 4.2: Overview of problem complexities

Programming implementation. In this case calculations are performed beginning with the smallest problems ($f(0)$ followed by $f(1)$, etc.) and continued iteratively. The interim results are stored in a table which allows the calculation algorithm to access previous calculation results. Using this approach, the calculation of $f(n)$ can be executed in both linear space and time.[169]

Building on the Dynamic Programming paradigm a wide variety of different (combinatorial) optimization problems can be solved with good real world performance. An example, where this principle is applied is the design of pseudo-polynomial algorithms for *NP*-complete problems, all of which are based on Dynamic Programming[170]. An example specifically discussed in this thesis is the weakly *NP*-complete Binary Knapsack Problem, which was mentioned in Chapter 4.1.1 and will be discussed in detail in Chapter 4.3.1.

## Approximating Optimal Solutions

While the calculation of an exact or optimal solution for a combinatorial optimization problem is the most desirable result, it is however not always feasible to calculate due to e.g. the problem size, available time or resources. This is particularly true when facing intractable problems such as the *NP*-complete problems discussed in Chapter 4.1.1[171]. Since a "good" solution is often acceptable in practical applications the following question presents itself: "If an optimization problem does not admit any efficient algorithm computing an optimal solution, is there a possibility to efficiently compute at least an approximation of the optimal solution?"[172]. Thus considerable research has gone into developing methods for the calculation of solutions, which are reasonably close to the exact optimum. This Chapter provides a brief overview of the subject.

---

[169]Cf. SKIENA (2008) The algorithm design manual, p. 273.
[170]Cf. VAZIRANI (2001) Approximation algorithms, p. 69.
[171]Cf. SCHUURMAN AND WOEGINGER (2007) Approximation Schemes, p. 2.
[172]HROMKOVIČ (2004) Algorithmics for Hard Problems: Introduction to Combinatorial Optimization, Randomization, Approximation, and Heuristics, p. 247.

## Greedy Algorithm

When discussing the approximation of optimal solutions, it is important to discus the **Greedy Algorithm** as a simple heuristic[173] to find a more-or-less good solution[174]. The underlying idea can be described in a single sentence: in every step of execution, it selects the alternative, which at that point in time appears to be the best. To provide a practical example the following combinatorial problem is briefly introduced: given a bounded set of coins with varying denomination the objective is to make change for a given amount of currency using a minimal number of coins. A Greedy algorithm would attempt to find the solution by iteratively adding the remaining largest coin to the solution set, where an addition of said coin would not cause the solution to exceed the target amount[175].

Unfortunately the algorithm is as simple as it is short-sighted[176]. In the example described above it is neither guaranteed that an acceptable solution will be found even if one exists nor is it certain, that a generated solution is optimal or even good. The algorithm searches for local optimums and will only "stumble upon" the global optimum by chance. Generally the success of searching for the optimal solutions with the Greedy algorithm depends heavily on the problem structure. In fact certain situations or problems do exist, where a Greedy algorithm yields optimal results. Examples hereof are the class of combinatorial problems referred to as Matroids, the generation of prefixes for the Huffman Code[177] or the Fractional Knapsack Problem [178].

In spite of the shortcomings discussed above, the Greedy algorithm being simple and straightforward. It can readily be adapted to a specific problem, is easy to implement and is very often time efficient[179]. Furthermore it can be used in conjunction with LP-relaxation to accurately calculate an upper bound for the Branch-and-Bound algorithm (refer to Chapter 4.1.2.1). Thus in summary it can be of considerable value when solving optimization problems.

---

[173]A "heuristic is a rule or solution adopted to reduce the complexity of computational tasks, thereby reducing demands on resources such as time, memory and attention." (cf. AUDI (1999) The Cambridge dictionary of philosophy, p. 379) or "rules of thumb that are used to find solutions to problems quickly" (KYNN (2008) Journal of the Royal Statistical Society: Series A, p. 242).

[174]Improving tractability at the cost of optimality is in general a characteristic of heuristic approaches(cf. ROTHLAUF (2011) Design of modern heuristics: Principles and application, p. 88).

[175]Cf. NEAPOLITAN (2015) Foundations of algorithms, pp. 151.

[176]Cf. JUNGNICKEL (2005) Graphs, networks and algorithms, p. 127.

[177]Cf. CORMEN (2009) Introduction To Algorithms, pp. 437.

[178]Cf. DANTZIG (1957) Operations Research, p. 266.

[179]Cf. PANDEY (2008) Design Analysis and Algorithm, p. 287.

## Approximation Schemes

As described above the main disadvantage of Greedy based approaches to approximate the optimal solution of an optimization problem is the absence of any assurances regarding the "goodness" of the generated solution. **Approximation algorithms** aim to provide a solution for problems with a certain minimum quality and a bounded time-complexity. The minimum stipulated solution goodness, which simultaneously described the algorithm's worst-case performance is denoted by the so-called *approximation ratio* denoted by $\rho$ is defined as

$$\rho \geq \max(\frac{f(x^{approx})}{f(x^*)}, \frac{f(x^*)}{f(x^{approx})}) \tag{4.4}$$

with $x^{approx}$ as the approximated solution and $x^*$[180]. Generally the resulting time-complexity $C$ depends on both the approximation ratio and the input-size $n$[181]. Approximation algorithms for difficult e.g. NP-complete problems, which achieve approximation ratios bounded by $1 + \epsilon$ are hence referred to as **approximation schemes**[182].

While there are many types of approximation schemes, the two most important in the context of this thesis are now briefly presented. The **fully polynomial-time approximation scheme** (**FPTAS**) is capable of calculating an approximate of the optimal solution with polynomially bounded $C$ both in $n$ and $1/\epsilon$. More specifically there exists a polynomial $p : \mathbb{R} \times \mathbb{R} \to \mathbb{R}$ for all $\epsilon > 0$ and all $n \in \mathbb{N}$ where $C$ is less than or equal to $p(\frac{1}{\epsilon}, n)$.[183] Thus if an FPTAS exists for a given combinatorial optimization problem exists, this eases the solving of real world problems. The **polynomial-time approximation scheme** (**PTAS**), while similar, differs in one important aspect: FPTAS $C$ is bounded by a poynomial in both $n$ and $1/\epsilon$, $C$ for an PTAS only needs to be bounded by a polynomial in $n$ and may feature $1/\epsilon$ as an exponent.[184] In practice this can result in rendering the running times polynomial in practice[185]. Concerning the possible existence of FPTAS algorithms for strongly NP-complete problems VAZIRANI shows that this is highly unlikely (i.e. only in the case should $N$ be equal to $NP$). Furthermore VAZIRANI demonstrates that with very weak restrictions all problems with an FPTAS must also admit a pseudo-polynomial time algorithm.[186]

[180]Cf. ROTHLAUF (2011) Design of modern heuristics: Principles and application, p. 88.
[181]Cf. SHACHNAI AND TAMIR Polynomial-Time Approximation Schemes, p. 9-1.
[182]Cf. ROTHLAUF (2011) Design of modern heuristics: Principles and application, p. 89.
[183]Cf. JANSEN AND MARGRAF (2008) Approximative Algorithmen und Nichtapproximierbarkeit, p. 177.
[184]Cf. ROTHLAUF (2011) Design of modern heuristics: Principles and application, p. 89.
[185]Cf. SHACHNAI AND TAMIR Polynomial-Time Approximation Schemes, p. 9-2.
[186]Cf. VAZIRANI (2001) Approximation algorithms, p. 71.

# Sequence and String Alignment

Sequence or string alignment originated in the area of computer science and today has applications in a wide array of other fields such as molecular biology[187], natural-language-processing[188], marketing research[189] and even sociology[190]. The superordinate objective is the comparison of one or multiple sequences and the assessment of their similarity. In biological research, a frequent application is the detection of similarity in DNA-strands[191] to infer common ancestry, by abstracting away the three dimensional nature of the DNA molecule[192]. This principle of abstracting a problem from a specific domain and simplifying it into a problem relating to the comparison of two sequences also applies to the other aforementioned fields. In computer science the sequence alignment problems are often referred to as string alignment or inexact matching problems. Here the similarity of two or more strings[193] is assessed in lieu of e.g. DNA sequences. This Chapter provides an overview of the research into sequence and string alignment problems.

## Hamming Distance

Among the first work in this field was the definition of the *Hamming Distance* by the American mathematician RICHARD HAMMING, as a measure for the similarity of strings of identical length. More specifically the Hamming distance is the number of non-identical symbols in the two strings[194]. While HAMMING's publication did not define the means for its calculation, computation in practice is trivial. When applied to binary codewords[195] calculation is particularly easy since a mere XOR[196] operation on the codewords, followed by counting the occurrences of the number 1 in the result is required. The Hamming Distance was the basis for work in the areas of coding theory for data transmission and ultimately error correcting codes.

---

[187]Cf. SHARMA (2009) Bioinformatics: Sequence Alignment and Markov Models, p. 41.
[188]Cf. CROCHEMORE, HANCART AND LECROQ (2007) Algorithms on Strings, p. vii.
[189]Cf. PRINZIE AND VAN DEN POEL (2006) Decision Support Systems, pp. 508.
[190]Cf. ABBOTT AND HRYCAK (1990) American Journal of Sociology, pp. 144.
[191]Deoxyribonucleic acid, is a molecule usually stored in the nucleus of biological cells, which contains the hereditary information of the respective organism. When stored in the nucleus of a cell, DNA exists helical structure comprised of two twisted strands.
[192]Cf. GUSFIELD (2009) Algorithms on Strings, Trees, and Sequences: Computer Science and Computational Biology, p. xiii.
[193]A string in this context is a linear sequence of concatenated symbols taken from an alphabet A
[194]Cf. HAMMING (1950) Bell System Technical Journal, pp. 147.
[195]i.e. string comprised of the symbols 1 and/or 0
[196]Exclusive-Or is a binary operation, which returns 1 if and only if exactly one of its operands is equal to 1 while the other remains 0

# Longest Common Substring Problem

Another basic problem in computer science is the determination of the longest common substring contained in two strings $S_1$ and $S_2$ with potentially non-equal lengths. A substring in this context is the longest contiguous series of characters contained in both strings. Given the two strings "Tactpallet" and "apalling" the longest common substring is "pall". In the field of bioinformatics this measure is used to identify similar regions between sequences and thus serves as a similarity metric.[197] The intuitive brute force solution to this problem is to simply compare the first character in $S_1$ to the first character in $S_2$ and to continue comparing the consecutive characters in both strings as long as they match. When a non matching character is encountered move on to the next character in $S_2$, repeat and finally move to the next character in $S_1$.[198] While this solution is trivial, its running time is cubic. To reduce running time approaches based on dynamic programming were applied to render time complexity quadratic. In 1997 the American mathematician Donald Knuth published the Knuth-Morris-Pratt algorithm capable of computing the longest common substring in linear time[199].

# Longest Common Subsequence Problem

Another classic computer science problem, which is quite closely related to the Longest Common Substring Problem, deals with the identification of the longest non-contiguous sequence of characters in the same relative order shared by two strings $S_1$ and $S_2$. Using the same example strings "Tactpallet" and "apalling" (chosen in Chapter 4.2.2), the longest common subsequence is "apall". Aside from applications in bioinformatics this measure of string similarity is used in word-processors and programming utilities to find similarities and changes in electronic documents edited by different users[200].

A brute force approach to solving this problem is conceptually trivial: after generating sets of every conceivable non-contiguous subsequence for $S_1$ and $S_2$ respectively, the longest sequence the intersection of both sets would could be selected. For a string with $n$ characters the number of possible subsequences is $2^n$. Thus this approach's drawback is its exponential time complexity rendering it infeasible for long strings. Since this problem exhibits an

---

[197]Cf. SUNG (2009) Algorithms in Bioinformatics: A Practical Introduction, p. 60.
[198]Cf. MCMILLAN (2007) Data Structures and Algorithms Using C#, p. 319.
[199]Cf. KNUTH, MORRIS, JR., JAMES H. AND PRATT (1977) SIAM Journal on Computing, pp. s323.
[200]Cf. SALOMON AND MOTTA (2010) Handbook of data compression, p. 1170.

optimal substructure (refer to Chapter 4.1.2.2) it can be solved using a Dynamic Programming algorithm in polynomial time.[201]

## Levenshtein Distance

In 1965 the Russian mathematician VLADIMIR LEVENSHTEIN expanded upon HAMMING's work by presenting a metric to compute a minimum number of edit operations necessary to transform one binary codeword into another. This "distance" between the two words is defined as *Levenshtein Distance* ($LD$) and sometimes also referred to as the bqiEdit Distance. With $\xi$ representing the empty word, edit operations in this context can be substitutions ($0 \mapsto 1$ or $1 \mapsto 0$), insertions ($\xi \mapsto 1$, $\xi \mapsto 0$) or deletions ($1 \mapsto \xi$, $0 \mapsto \xi$)[202]. While the metric's definition was first published in 1966, no algorithm for its calculation was included. The American researcher FREDERICK DAMERAU presented an expansion on the Levenshtein Distance metric which adds the transposition operation of two adjacent characters to the set of operations introduced by LEVENSHTEIN[203]. Since this extension however offers only little tangible added value but significant complexity the focus of this Chapter will be LEVENSHTEIN's definition.

In the years after LEVENSHTEIN's publication of the similarity metric, various algorithms to perform the non-trivial computation were developed[204]. FISHER AND WAGNER'S variant offers simplicity combined with sufficient performance regarding computation time and space[205]. Thus this algorithm is chosen for the purposes of this thesis and is described below.

When calculating Levenshtein Distance between the strings $S_1$ and $S_2$, the permissible insertion, deletion and substitution operations will henceforth be denoted as $i,d$ and $s$. While they do not represent edit operations, matches found in the sequence will be represented by the symbol $m$. The sequence of operations $e$ necessary to perform the transformation is defined as the edit transcript, where each element is an element of the alphabet $i,d,s,m$. The Levenshtein distance $\|e_{opt}(S_1, S_2)\|$ is defined as the minimum number of edit operations required to transform $S_1$ into $S_2$. The corresponding transcript containing the minimum number of edit

[201]Cf. LI, LOOI AND ZHONG (2006) Advances in intelligent IT: Active media technology 2006, p. 244.
[202]Cf. LEVENSHTEIN (1966) Soviet Physics Doklady, p. 707.
[203]Cf. DAMERAU (1964) Communications of the ACM, p. 171.
[204]Cf. FISCHER AND WAGNER (1974) String-to-String Correction Problem, pp. 168; MAJSTER AND REISER (1980) SIAM Journal on Computing, pp. 785; TICHY (1984) ACM Transactions on Computer Systems, pp. 309.
[205]With $S_1$ and $S_2$ as the input strings to compare, the required computational time and memory both increase as functions of $|S_1| \cdot |S_2|$

operations is the optimal transcript $e_{opt}(S_1, S_2)$. If more than one optimal transcript exists, these are referred to as being "cooptimal"[206].

The following approach to computing the Levenshtein Distance is based on the dynamic programming paradigm, where problems are broken down into subproblems, whose solutions are computed and stored for future retrieval without recomputation[207]. When applying this to the problem of calculating the Edit distance between $S_1$ and $S_2$, the distance between all substrings $S_1[1..i]$ and $S_2[1..j]$ with $i, j \in \mathbb{N}$, $0 \leq i \leq |S_1|$ and $0 \leq j \leq |S_2|$ are calculated and stored in a Matrix $D$ with the dimensions of $(|S_1| + 1) \times (|S_2| + 1)$. After initializing both the top and left borders of the matrix with

$$D(i, 0) = i \tag{4.5}$$

and

$$D(0, j) = j \tag{4.6}$$

the remaining entries $D[i, j]$ are calculated iteratively from those already contained in $D$ based on the recurrence relation

$$D(i, j) = min \begin{cases} D(i - 1, j) + 1 \\ D(i, j - 1) + 1 \\ D(i - 1, j - 1) + eq(i, j) \end{cases} \tag{4.7}$$

where

$$eq(i, j) = \begin{cases} 1 : S_1(i) \neq S_2(j) \\ 0 : S_1(i) = S_2(j) \end{cases}. \tag{4.8}$$

The calculation of the full strings' Edit Distance is complete, when the bottom right entry $D(|S_1|, |S_2|)$, which then contains $\|e_{opt}(S_1, S_2)\|$ has been calculated. A valid transcript $e_{opt}$ can now be derived from the matrix $D$ by using a traceback algorithm, which finds a path from $D(|S_1|, |S_2|)$ to $D(0, 0)$ with $\|e_{opt}(S_1, S_2)\|$ edit operations.

In the algorithm described by GUSFIELD movement between entries in $D$ can either be upwards vertically, leftwards horizontally or simultaneously up- and leftwards diagonally.

---

[206]Cf. GUSFIELD (2009) Algorithms on Strings, Trees, and Sequences: Computer Science and Computational Biology, p. 216.
[207]Cf. POOLE AND MACKWORTH (2010) Artificial Intelligence: Foundations of Computational Agents, p. 103.

While a vertical movement represents an insertion operation, horizontal movement represents the complementary deletion operation. If $S_1(i) \neq S_2(j)$ diagonal movement represents a substitution or otherwise no operation. With $D(i,j)$ as the current entry movement direction is chosen as

$$\uparrow \text{ if } D(i,j) = D(i-1,j) + 1 \tag{4.9}$$

$$\nwarrow \text{ if } D(i,j) = D(i-1,j-1) + 1 \text{ and } S_1(i) \neq S_2(j) \tag{4.10}$$

$$\nwarrow \text{ if } D(i,j) = D(i-1,j-1) \text{ and } S_1(i) = S_2(j) \tag{4.11}$$

$$\leftarrow \text{ if } D(i,j) = D(i,j-1) + 1. \tag{4.12}$$

If more than one of these apply, two or more cooptimal transcripts exist and the pursuance of any of the candidate branches will lead to an optimal transcript.[208]

# Knapsack Problems

The Knapsack Problem is considered one of the classical discrete programming problems from the area of combinatorial optimization[209]. Although it was formally first referred to as such by GEORGE DANTZIG in 1957[210], the problem was already discussed in publications as early as 1897[211]. Also in 1957 BELLMAN explored the applicability of Dynamic Programming (refer to Chapter 4.1.2.2) as a means of solving the Knapsack Problem[212]. Several years later in 1972 Richard Karp proved that the problem is NP-complete[213].[214] After introducing the classical binary Knapsack Problem, this Chapter outlines several derivatives thereof, which were developed in the ensuing years.

[208]Cf. GUSFIELD (2009) Algorithms on Strings, Trees, and Sequences: Computer Science and Computational Biology, pp. 217.
[209]Cf. MARTELLO AND TOTH (1990) Knapsack problems: Algorithms and computer implementations, p. xi.
[210]Cf. DANTZIG (1957) Operations Research, pp. 266.
[211]Cf. MATHEWS (1897) Proceedings of the London Mathematical Society, pp. 486.
[212]Cf. BELLMAN (1957) Operations Research, pp. 723.
[213]Cf. KARP Reducibility among Combinatorial Problems, p. 95.
[214]Cf. KELLERER, PFERSCHY AND PISINGER (2004) Knapsack Problems, p. III.

## The Original Binary Knapsack Problem

The basic Knapsack Problem is known as the **Binary** or *0-1 Knapsack* ($01KP$). Due to its wide practical applicability, this is the most important variant and has been the subject of extensive research[215]. The basic premise of the problem is as follows:

When packing his supplies for an extended journey a mountaineer must choose wisely. He can choose supplies from a set of $n$ items, each of which has a specific value $v_i$ and weight $w_i$. As a means of transportation the mountaineer would like to use his eponymous knapsack with a limited carrying capacity $c$. Driven by his intrinsic desire to survive, the mountaineer would like to maximize the value of all items he is able to take on his journey. In its most basic form the Binary Knapsack Problem additionally requires the mountaineer to make a binary decision of whether or not to place an item into his Knapsack. Thus packing items partially is not permitted.[216]

Thus given a set of $n$ items and a knapsack $k$

$v_i$   value of the $i$th item

$w_i$   weight of the $i$th item

$c$    maximum carrying capacity of the knapsack

the problem is formally defined as

$$\text{maximize} \quad p = \sum_{i=1}^{n} v_i x_i \tag{4.13}$$

$$\text{subject to} \quad \sum_{i=1}^{n} w_i x_i \leq c \tag{4.14}$$

$$x_i \in \{0, 1\}, \quad i \in N = \{1, ..., n\} \tag{4.15}$$

where

$$x_i = \begin{cases} 1 & \text{if item i is placed in knapsack} \\ 0 & \text{else.} \end{cases} \tag{4.16}$$

As mentioned in the introduction above, this weakly *NP*-complete form of the problem has been the focus of extensive research in the last sixty years. Based on BELLMAN'S Dynamic Programming paradigm several algorithms to calculate an exact solution to the problem in pseudo-polynomial time were introduced in the 1950s. In 1957 DANTZIG proposed a

---

[215]Cf. MARTELLO AND TOTH (1990) Knapsack problems: Algorithms and computer implementations, p. 13.
[216]Cf. CORMEN (2009) Introduction To Algorithms, pp. 425

relaxation of the binary restriction $x_i \in \{0, 1\}$ as a means of calculating the upper bound for possible solutions. This idea proved to be a strong influence on Knapsack Problem-research for many years, until MARTELLO and TOTH developed a superior method of determining an upper bound[217]. Approximately ten years later in 1967 KOLESAR presented a Branch and Bound-based (refer to Chapter 4.1.2.1) approach to the solving the problem[218]. In the following years KOLESAR'S research focused on improvements and expansions to his work.[219] In 1975 IBARRA presented the first fully polynomial-time approximation scheme (refer to Chapter 4.1.3.2) for the Knapsack and Subset of Sums Problems[220]. Since then research regarding Knapsack Problems has focused on improving the performance when handling large problems, the development of new algorithms for calculation of exact solutions, heuristics and approximation schemes[221].[222]

## Derived Knapsack Problem Types

While the various approaches to solving the Binary Knapsack Problem have been the focus of research for many years, the extension of the original problem and derivation of new problem types has also been the object of intensive academic discourse. This Chapter will provide an overview of the various variations developed and discussed over the years. table 4.1 at the end of this Chapter provides an overview of all the discussed problems and their respective properties.

Substituting the "less than or equal" relational operator with an "equals" yields the arguably closest relative of 01KP the *Equality Knapsack Problem* $(EKP)$. Thus put simply, the objective is to pack a set of items, which have a total weight exactly equal to the knapsack's capacity. As the 01KP it was derived from, the EKP is considered to be *NP*-hard. While various Branch-and-Bound and Core[223] based algorithms to solve the problem exactly have been presented, the problem still remains difficult to solve in certain scenarios[224].

---

[217]Cf. MARTELLO AND TOTH (1977) European Journal of Operational Research, p. 169.

[218]Cf. KOLESAR (1967) Management Science, p. 723.

[219]Cf. MARTELLO AND TOTH (1990) Knapsack problems: Algorithms and computer implementations, p. 14.

[220]Cf. IBARRA AND KIM (1975) J. ACM, p. 463.

[221]Cf. KELLERER, PFERSCHY AND PISINGER (2004) Knapsack Problems, p. III.

[222]Cf. MARTELLO AND TOTH (1990) Knapsack problems: Algorithms and computer implementations, p. 14.

[223]The Core Concept was introduced by BALAS in 1980 and builds on the idea that when solving knapsack problems it is only difficult to decide to include an item in the solution for a subset of all items. Thus computational effort is focused on this subset of the variables (i.e. items) which is referred to as the problem core. For further information please refer to BALAS AND ZEMEL (1980) Operations Research.

[224]Cf. AARDAL AND LENSTRA Hard Equality Constrained Integer Knapsacks, p. 350; VOLGENANT AND MARSMAN (1998) The Journal of the Operational Research Society, p. 86; RAM AND SARIN (1988) The Journal of the Operational Research Society, p. 1045.

Another close relative of the 01KP is the **Minimization Knapsack Problem** ($MinKP$). As implied by the name the objective is to minimize the value of the packed item subject to the weight of all packed items being greater than or equal to a minimum capacity[225]. Minimizing the value of the items packed into the knapsack under the antecedent condition is equivalent to maximizing the value of the remaining items. Thus every MinKP can easily be transformed into a 01KP and solving the problem precisely is trivial. The approximation algorithms however cannot simply be applied to the transformed problem. This gives rise to research focusing on the approximation of the optimal solutions over the years [226].

Two common Knapsack Problem subtypes which were introduced in the 1960s are the **Bounded Knapsack Problems** ($BKP$) and the **Unbounded Knapsack Problems** ($UKP$). The problem modification in this case focuses on the items available for packing. While the original problem statement referenced $n$ individual items with defined weights and values, the bounded and unbounded refer to classes of items with these properties. The difference between BKP and UKP lies within the limit of items contained in the item classes. On the one hand BKPs have limited numbers of items per class as the name implies. In principle a transformation from BKP to Binary Knapsack problem can be performed trivially by creating individual items for each item in the available item classes. The problem however with this approach lies in the resulting total problem size, which can cause skyrocketing complexity when high limits exist per problem class. UKPs contain an unlimited number of item instances per class. Thus the aforementioned trivial transformation does not apply. Research for both problem categories has yielded specialized algorithms based on the Dynamic Programming and Branch and Bound principles. Additionally FPTAS were presented both for the BKS and UKS by KELLERER and IBARRA respectively.[227]

Another well-studied NP-hard problem variant with many practical applications is the **d-Dimensional Knapsack Problem** ($dKP$)[228]. In real-world applications, both the weight and weight-constraints of a knapsack problem can be subject to multiple constraints and thus be multi-dimensional. In the scenario of the mountaineer introduced in Chapter 4.3.1 the restricted volume of the knapsack is not reflected in the problem definition. By giving each of the $n$ items a volume constraint in addition to the physical weight constraint, the original problem could be extended to a 2-Dimensional Knapsack Problem. The copious

[225] Cf. DIUBIN AND KORBUT (2011) Journal of Computer and Systems Sciences International, pp. 182.
[226] Cf. GENS, GEORGII V. AND EUGENII V. LEVNER (1979) Mathematical Foundations of Computer Science, pp. 292; GÜNTZER AND JUNGNICKEL (2000) Operations Research Letters, pp. 292.
[227] Cf. KELLERER, PFERSCHY AND PISINGER (2004) Knapsack Problems, p. 185.
[228] Cf. Ibid., p. 235.

work in this area has yielded various Integer Linear Programming based, metaheuristic and collaborative approaches, which ease the solving of dKPs[229]. While many heuristics exist, unlike the previously discussed problem variants, it has been shown that dKP is strongly $NP$-hard and thus can only be approximated with PTAS[230]. Solving problems instances with large numbers of dimensions remains time-consuming[231].

Assuming the mountaineer can carry more than one knapsack on his journey, this yields a new problem variant referred to as the ***Multiple Knapsack Problem*** $(MKP)$. The differentiating feature of the MKP is the introduction of multiple knapsacks, each of which has a defined maximum capacity. Items can be assigned to a maximum of one knapsack. In practice the MKP has a wide range of practical applications from logistical optimization to financial portfolio optimization[232]. Since the MKP is a strongly $NP$-hard problem variant, it has has been shown, that even with only two knapsacks, no FPTAS can exist. A PTAS was however introduced by CHEKURI and KHANNA.[233]

From its basic premise of considering classes of items the ***Multiple Choice Knapsack Problem*** $(MCKP)$ has some similarity with the BKP and UKP presented above. Here the original binary knapsack problem is extended by adding disjoined multiple-choice constraints within each of the aforementioned item classes[234]. This variation of the problem can be used, when the choice of one item within a class precludes the choice of another item within the same class. When applied to the example the mountaineer might need to take a toothbrush with him on his journey but has little use for more than one brush. Should he however have multiple toothbrushes to choose from, defining a toothbrush class containing all options could be used to model this decision problem. Aside from the universally applicable Dynamic Programming and Branch and Bound based solutions, FPTAS have been presented for the MCKP by CHANDRA and LAWLER. [235]

Introduced by GALLO in 1980, that ***Quadratic Knapsack Problem*** $(QKP)$ deals with situations, in which the value of an item depends on the other items also included in the knapsack[236]. This is a reflection of many real life applications in which the utility or value

[229]Cf. PUCHINGER, RAIDL AND PFERSCHY (2010) INFORMS Journal on Computing, p. 1.
[230]Cf. KULIK AND SHACHNAI (2010) Information Processing Letters, p. 708.
[231]Cf. FRÉVILLE (2004) European Journal of Operational Research, p. 1.
[232]Cf. (2000)Intelligent Problem Solving. Methodologies and Approaches, p. 296.
[233]Cf. CHEKURI AND KHANNA (2005) SIAM Journal on Computing, p. 713.
[234]Cf. PISINGER (1995) European Journal of Operational Research, p. 394.
[235]Cf. CHANDRA, HIRSCHBERG AND WONG (1976) Theoretical Computer Science, pp. 292; LAWLER (1979) Mathematics of Operations Research, pp. 205.
[236]Cf. GALLO, HAMMER AND SIMEONE Quadratic knapsack problems, p. 132.

of an item reflects how well the given selection of items fits together[237]. When applied to the mountaineer's situation, this variation of the problem might reflect the fact, that e.g. packing coffee without a proper pot to heat it in would be of little value. Since the QKP's introduction it has been the objective of intensive research[238]. There are many resulting publications on the determination of upper bounds through various forms of constraint relaxation. In 2002 RADER and WOEGINGER presented a Dynamic Programming algorithm with pseudo-polynomial time complexity and an FPTAS for the special case where the underlying graph is edge series-parallel[239].

Many problems require finding solutions, which are optimal in terms of more than one objective function. This gave rise to the definition of the *Multiobjective Knapsack Problem* $(MOKP)$. When applied to the knapsack packed by the mountaineer each item could contribute to the upcoming journey in a different fashion. Examples of possible value dimensions for supplies in this case are: nutritional value, hydrational value, protective value etc.. By thus assigning a vector of values to each item, the optimization can be performed in light of different objectives. Over the years various approaches to calculate exact solutions to the problem based on the Dynamic Programming paradigm[240]. In 2002 ERLEBACH, KELLERER and PFERSCHY presented both a FPTAS for the MOKP and a PTAS for the m-dimensional MOKP[241].

The *Precedence Constraint Knapsack Problem* $(PCKP)$ expands on the Binary Knapsack problem by additionally taking cases into account in which the addition of an item is contingent on packing of another item. In literature these so-called *precedence relationships* are often represented by a directed acyclic graph[242]. Applied to the mountaineer an example of this would be the necessity to pack a can-opener prior to packing canned food. Practical real world applications range from network design[243] to production planning[244]. The PCKP was first formally introduced in 1983 by JOHNSON and NEIMI, who proved it to be strongly NP-hard and presented an algorithm to calculate the exact results[245]. In 1998 SHAW and CHO presented a Branch and Bound based variant[246]. In the more recent past YOU and

[237]Cf. KELLERER, PFERSCHY AND PISINGER (2004) Knapsack Problems, p. 348.
[238]Cf. PISINGER (2007) Discrete Applied Mathematics, p. 623.
[239]Cf. RADER JR. AND WOEGINGER (2002) Operations Research Letters, pp. 159.
[240]Cf. HUNG AND FISK (1978) Naval Research Logistics Quarterly, pp. 571; EUGÉNIA CAPTIVO et al. (2003) Computers & Operations Research, pp. 1865.
[241]Cf. ERLEBACH, KELLERER AND PFERSCHY (2002) Management Science, pp. 1603.
[242]Cf. BOLAND et al. (2012) Mathematical Programming, p. 481.
[243]Cf. SHAW, CHO AND CHANG (1997) Telecommunication Systems, pp. 29.
[244]Cf. MORENO, ESPINOZA AND GOYCOOLEA (2010) Electronic Notes in Discrete Mathematics, pp. 407.
[245]Cf. JOHNSON AND NIEMI (1983) Mathematics of Operations Research, p. 1.
[246]Cf. SHAW AND CHO (1998) Networks, p. 205.

YAMADA published an approach relying on Langrangian relaxation and were able to solve problems with thousands of items within a few minutes on an ordinary workstation[247]. Due to its strongly *NP*-hard nature, a general FPTAS has not been found to date. In 2007 however KOLLIOPOULOUS found an FPTAS for the important case of a given two-dimensional partial order[248].

Formally presented YU in 1997 the ***Min-Max 0-1 Knapsack Problem*** ($MMKP$) is among the younger knapsack problems discussed in this thesis[249]. The original problem is extended by introducing a multi-dimensional value vector per item as is the case in MOKP. The resulting value of a solution is the sum of all item values and thus also a vector. The optimization objective in the MMKP is to find the combination of items, which maximizes the minimum of all solution value vector dimensions and thus objectives. An example-application of the MMKP to the situation the mountaineer finds himself in would be the planning for different weather scenarios - each being represented by one dimension in the value vector. Since the different items could have different values depending on the scenario - e.g. the mountaineer might have little use for snow shoes if he does not encounter freezing weather - it might be beneficial to find the solution with the largest total value in the worst case scenario. Thus the minimum value found in any of the total value's dimensions is minimized. Literature describes practical applications in the areas of financial investment planning and military logistics[250]. Due to its relative novelty, research on this problem is comparatively limited. Regarding the finding of exact solutions work has focused on Branch-and-Bound based algorithms[251]. Since YU demonstrated that the problem is strongly *NP*-hard and thus finding an FPTAS is highly unlikely.

In certain situations the capacity available in the knapsack might depend on the composition of the items in the knapsack. A simple case is the dependency on the number of packed items, where each item brings with it a constant additional weight[252]. While this simplification eases explanation of the problem other functions describing the dependency of additional weight and number of packed items are possible[253]. When the problem was first introduced in 1978 it was formulated to model the behavior of a communications band share by means of the

---

[247]Cf. YOU AND YAMADA (2007) European Journal of Operational Research, p. 618.
[248]Cf. KOLLIOPOULOS AND STEINER (2007) Discrete Applied Mathematics, p. 889.
[249]Cf. YU (1996) Operations Research, p. 407.
[250]Cf. Ibid., pp. 407; KRESS, PENN AND POLUKAROV (2007) Naval Research Logistics, pp. 656.
[251]Cf. PINTO et al. (2015) Mathematical Problems in Engineering, p. 1; TANIGUCHI, YAMADA AND KATAOKA (2008) Computers & Operations Research, p. 2034; IIDA (1999) Journal of Combinatorial Optimization, p. 89.
[252]Cf. KELLERER, PFERSCHY AND PISINGER (2004) Knapsack Problems, p. 416.
[253]Cf. PFERSCHY, PISINGER AND WOEGINGER (1997) Discrete Applied Mathematics, pp. 271.

FDMA[254]. Inherent in this form of multiplexing approach is the fact that the number of simultaneous users imposes an overhead which must be subtracted from the total available capacity. Aside from this example, other practical applications include the packing of fragile items on a truck and timesharing on computer systems and even aspects of shopping center design[255]. Generally, problems where the number of items causes knapsack capacity to decrease are referred to as *Collapsing Knapsack Problems* ($CoKP$). Problems where the opposite is the case are referred to as *Expanding Knapsack Problems* ($ExKP$). An example of an application lies in the area of purchasing, where sales-volume based discounts are common. In 1997 PFERSCHY showed, that every instance of CoKP can be reformulated as a 01KP with $2n$ elements.[256] Thus in principle CoKP can be solved exactly using the same methods. Nevertheless research on the CoKP has yielded specialized exact solutions in addition to CPK-specific approximating approaches[257].

The newest derivative of the 01KP reviewed in the context of this paper is the *Integer Knapsack Problem with Setup Weights* ($IKPSW$), which was introduced by MCLAY in 2007. The underlying idea of this problem variant is that the weight of an item added to the knapsack can depend on the items already in or, better, not yet in the knapsack. Thus it bears a similarity to the CoKP and ExKP presented above. Unlike these problems however the weight of an item is not a function of the number of items in the knapsack, but rather depends on the types of other items contained in the knapsack. More specifically, adding an item causes the solution to incur a "setup cost" if no other items belonging to the added items "family" are already in the knapsack. When applied to the situation the mountaineer decision problem, certain items might have compatible shapes, where selecting two complementary items for packing might serve to reduce the required volume below the sum of the two items' individual volumes. Originally the problem was formulated to optimize the passenger screening at airports, but it has many other applications such as economics or production scheduling. In her introductory paper MCLAY provided and FPTAS for the IKPSW, several years later in 2015

---

[254]Frequency Division Multiple Access is an access protocol developed to allow several users to use a communications channel simultaneously by diving the available frequency range into multiple frequence bands. FDMA is common both in long-range (e.g. Satellites) and short-range (DECT cordless phones) communications. To allow for proper communications the individual frequency bands must be separated by gaps.

[255]Cf. POSNER AND GUIGNARD (1978) Mathematical Programming, p. 155.

[256]Cf. PFERSCHY, PISINGER AND WOEGINGER (1997) Discrete Applied Mathematics, pp. 271.

[257]Cf. FAYARD AND PLATEAU (1994) Discrete Applied Mathematics, pp. 175; WU AND SRIKANTHAN (2006) Information Sciences, pp. 1739.

CHEBIL presented a Dynamic Programming based algorithm, which calculates an optimal solution in pseudo-polynomial time.[258]

A derivative of the 01KP, which combines aspects of several other knapsack problem derivatives is the **Change-Making Problem** ($CMP$). Given a defined sum of money and unlimited coins in various denominations the basic premise of the CMP is to find a selection of bills and coins, which has a value equal to the defined amount while simultaneously minimizing the number of required coins. Thus the problem combines Unbounded Knapsack Problem, Minimization Knapsack Problem and the Equality Knapsack Problem as a reflection of the unlimited number of available coins, the minimization objective and the equality constraint. While intuitively the problem might appear to be solvable trivially, it was shown to be *NP*-hard by LUECKER[259]. This is due to many currencies such as the Dollar and Euro being so-called canonical systems[260]. Given a canonical currency, a greedy algorithm will always yield the optimal result[261]. CHANG and GILL provide deeper insight into the prerequisites of a greedy algorithms optimality[262]. To solve the general problem for an arbitrary coin systems MARTELLO and TOTH investigate the estimation of lower bounds for problem instances[263].

| Acr. | Problem Type | Modification | Complexity[264] |
|------|-------------|-------------|-----------|
| EKP | Equality Knapsack Problem | equality constraint for weight | w-$NP$-hard |
| MinKP | Minimization Knapsack Problem | minimization of value with miminal weight constraint | w-$NP$-hard |
| BKP | Bounded Knapsack Problem | item classes with limited item instances | w-$NP$-hard |
| UKP | Unbounded Knapsack Problem | item classes with unlimited item instances | w-$NP$-hard |
| dKP | d-Dimensional Knapsack Problem | multi-dimensional weights and constraints | s-$NP$-hard |

[258]Cf. MCLAY AND JACOBSON (2007) Computational Optimization and Applications, pp. 35; CHEBIL AND KHEMAKHEM (2015) Computers & Operations Research, pp. 40.

[259]Cf. LUEKER (1975) Two NP-complete problems in nonnegative integer programming, pp. 1.

[260]A complete coverage of a canonical coin system's properties does not lie within the scope of this thesis. For further information on the subject including an algorithm to test if a coin system is canonical please refer to PEARSON (1994) Operations Research Letters.

[261]Cf. CAI (2009) Canonical Coin Systems for CHANGE-MAKING Problems, p. 499.

[262]Cf. CHANG AND GILL (1970) Journal of the ACM, pp. 113.

[263]Cf. MARTELLO AND TOTH (1980) European Journal of Operational Research, p. 322.

| Acr. | Problem Type | Modification | Complexity[264] |
|---|---|---|---|
| MCKP | Multiple Choice Knapsack Problem | item classes with mutually exclusive multiple choice constraints in each class | s-$NP$-hard |
| QKP | Quadratic Knapsack Problem | item values depend on items already in knapsack | s-$NP$-hard |
| MOKP | Multiobjective Knapsack Problem | multi-dimensional value | s-$NP$-hard |
| PCKP | Precedence Constraint Knapsack Problem | adding of items contingent on items already in knapsack | s-$NP$-hard |
| MMKP | Min-Max 0-1 Knapsack Problem | calculation of solution which maximizes the minimal value in all value dimensions | s-$NP$-hard |
| CoKP | Collapsing Knapsack Problem | adding of items causes additional capacity overhead | w-$NP$-hard |
| ExKP | Expanding Knapsack Problem | adding of items provides additional capacity bonus | w-$NP$-hard |
| IKPSW | Integer Knapsack Problem with Setup Weights | adding of items triggers capacity bonus if item of same class already in knapsack | w-$NP$-hard |
| CMP | Change Making Problem | adding of items provides additional capacity bonus | w-$NP$-hard |

Table 4.1: Overview of 0-1 Knapsack Problem Derivatives

---

[264]The acronyms w and s are used for weakly and strongly respectively

# Chapter 5

# Concept

This Chapter provides a detailed description of the algorithms at heart of this thesis using the notation introduced in Chapter 2.4. To this end fundamental definitions and variables are first introduced. The subsequent Chapters then focus on describing the ensuing "Tactsequence Mapping" and "Production Scheduling" algorithms.

## Fundamental Definitions

Based on the synchronized manufacturing concept described in Chapter 2.3.3, necessary variables and abbreviations required in the subsequent Chapters will now be introduced.

The fundamental unit of time, which represents the time available for production is the **Tact** $TA$. Each Tact has an ID (TID), an equal duration (Duration) and is mapped to a point in time at which it begins (PIT).

$$TA = \{TID, PIT, Duration\} \tag{5.1}$$

The *Calendar* $Cal$ is a set of all defined Tacts, which begin at strictly monotonic increasing points in time. Thus

$\forall t, k \in Cal$ with $t.TID < k.TID$

$$\pi_{Duration}(t) = \pi_{Duration}(k) \tag{5.2}$$

and

$$\pi_{pit}(t) < \pi_{pit}(k) \tag{5.3}$$

On the factory floor, the actual manufacturing processes are performed at machines or manual workstations. Since machinery in the broader sense is involved in almost all cases, the station performing a manufacturing task is referred to as a *Machine* $MA$. Each Machine is identified by a unique identifier (MID).

$$MA = \{MID\} \tag{5.4}$$

The set of all Machines present on the factory floor is the *Machine Complement* $MC$. For every Tact in the Calendar, each machine must dispose of defined available processing capacities $CA$. In line with the ideas in ZISKOVEN'S work, the available *Processing Capacity* $CA$ is defined by a *target-capacity* (TargetCapacity) and *maximum-capacity* (MaximumCapacity).

$$CA = \{MID, TID, TargetCapacity, MaximumCapacity\} \tag{5.5}$$

As the name implies, target-capacity is the ideal amount of Processing Capacity to be used during one Tact. Ideality is in this case exogenously determined by the organization deploying the synchronized production system. Maximum-capacity is the absolute maximum amount of processing capacity available. With instruments such as overtime or labor leasing, maximum-capacity can exceed target-capacity. In practical experience a target- to maximum-capacity ratio of approximately 0.8 is quite common.

The full set of all available processing capacities is defined as the *Capacity Supply* $CS$.

$\forall t \in (MC \times Cal) \exists c \in CS$

$$(\pi_{MID}(c) = \pi_{MID}(t)) \wedge (\pi_{TID}(c) = \pi_{TID}(t)) \tag{5.6}$$

$\forall t \in CS$

$$0 \leq \pi_{target-capacity}(c) \leq \pi_{maximum-capacity}(c) \tag{5.7}$$

In the presented synchronized production system materials are combined into bundles and transported together on Tactpallets $TP$. Tactpallets travel along Tactlines $TL$, each of which have a unique identifier (TLID). The set of all Tactlines is defined as the *Tactline Container* $TLC$.

$$TL = \{TLID\} \tag{5.8}$$

A Tactline is an ordered sequence of **Tactstations** $TS$ to which a Tactpallet is transported before a Tact begins. At a Tactstation the processing of semi-finished components proceeds during the Tact. Each Tactstation in a Tactline has an integer as unique identifier (TSID), which encodes the relative position of a Tactstation in a Tactline. To simplify ensuing calculations in this thesis, the first Tactstation in a Tactline has the TSID 1 while the last has the TSID equal to the length of the respective Tactline. In addition to the contained identifier, each Tactstation references the ID of the Tactline within which it is contained (TLID). The full set of all Tactstations in all Tactlines is the **Tactstation Container** $TSC$.

$$TS = \{TSID, TLID\} \tag{5.9}$$

The actual processing is however not performed by the Tactstation, but rather by one of the Tactsubstations $TSS$ contained therein. In essence a Tactsubstation represents one of the machines on the shop-floor defined above. By allowing for multiple Tactsubstations within a Tactstation, the common scenario in which more than one machine of a given type exists is taken into account. Thus a Tactline could contain a turning Tactstation which could in turn consist of three turning machines or Tactsubstations. Generally a Tactsubstation features an identifier (TSSID) which is unique within the respective Tactstation. Furthermore it includes references to the enclosing Tactstation (TSID) and Tactline (TLID). All Tactsubstations contained in all Tactstations of all Tactlines are elements of the **Tactsubstation Container** $TSSC = \{TSS_1...TSS_p\}$.

$$TSS = \{TSSID, TSID, TLID, MID\} \tag{5.10}$$

While a machine can be contained in only one Tactsubstation of any given Tactstation, it can however be contained in more than one Tactstation of the same or different Tactlines(refer to Figure 5.1).

$$\forall t \in (TLC \bowtie TSC \bowtie TSSC) \exists m \in MC$$

$$\pi_{MID}(t) = \pi_{MID}(m) \tag{5.11}$$

$$\forall t \in \gamma_{TLID,TSID,TSSID,MID,Count(MID)}(TLC \bowtie TSC \bowtie TSSC)$$

$$\pi_{Count(MID)}(t) = 1 \tag{5.12}$$

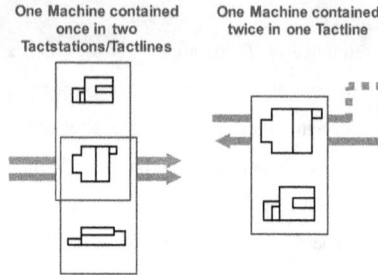

Figure 5.1: Machine in to multiple Tactlines and Machine in one Tacline multiple times

To simplify the ensuing description of the Tactsequence Mapping and Production Scheduling algorithms, a special type of Tactline is defined. Since generally Tactlines containing exactly one Tactstation would not represent a true flow of materials from one machine to another, these are henceforth referred to as Isolated Manufacturing Tactlines. The purpose of these is to allow Isolated Manufacturing Steps to be inserted between the manufacturing occurring in the context of a regular Tactline with multiple stations. E.g. with a factory floor containing $n$ machines, a practical Isolated Manufacturing Tactline could consist of exactly one Tactstation containing $n$ Tactsubstations. Each of these could represent one of the aforementioned $n$ machines. This setup would allow all machines on the shop floor to be used for manufacturing operations, which do not fit within predefined Tactlines.

A component scheduled for production is represented by a *Job* $J$, the complete set of which is contained in the *Job Pool* $JP$. A Job has a unique identifier (JID) and a specific *due-date* (DueDate), by which the manufacturing process must be completed at the latest. This date is generally calculated from deadlines in the superordinate tool creation project defined during the order acquisition step of the tool manufacturing process chain (refer to Chapter 2.2.2). Furthermore, each Job is directly linked to a Process Step Series and a Tactsequence, both of which are described below. During algorithm execution more than one loosely linked instance of both can exist for a particular Job. $PSSID$ on the one hand references the routing generated in Production Engineering after receiving Input-Data Processing. $TSeqID$ on the other hand links to the Tactsequence generated upon completion of Tactsequence Mapping.

$$J = \{JID, PSSID, TSeqID, DueDate\} \tag{5.13}$$

When production engineering completes the routing for a component, this routing is in essence an ordered set of **Production Steps** $PS$, which are necessary to successfully manufacture a product. An ordered set of Process Steps is defined as a **Process Step Series** $PSS$. The series constituting the routing is further qualified as $PSS_{routing}$. Each Process Step contains a link to the associated Job (JID), the required Machine capacity (RequiredCapacity), the relative order within the series (SIdx) and information regarding the required Machine setup (Rigging-Key). The latter can be used to describe which rigging activities must be completed on a Machine prior to commencing processing. By specifying a unique identifier, for e.g. a certain fixture, the Production Scheduling algorithm can later attempt to schedule multiple Jobs to be processed on a certain Machine in the same Tact. This reduces the number of necessary rigging operations and thus increases the proportion of productive time in a Tact. If multiple Machines are capable of executing a given manufacturing operation, multiple Process Steps exist, each of which encode one of the alternatives. During Input-Data Processing (refer to Chapter 5.2.1.2), $PSS_{routing}$ is transformed into $PSS$. The full complement of all Process Steps for all Jobs is contained in the **Process Step Container** $PSC$. The aforementioned $PSS$ for a specific Job $j$ is thus equal to $\sigma_{(JID=j.JID)\wedge(PSSID=j.PSSID)}(PSC)$ and represents the input Tactsequence Mapping (refer to Chapter 5.2).

$$PS = \{JID, PSSID, SIdx, MID, RiggingKey, RequiredCapacity\} \qquad (5.14)$$

After the completion of Tactsequence mapping, all Jobs, which can be processed by the synchronized production system have been assigned a Tactsequence $TSeq$ corresponding to their $PSS$. A Tactsequence is comprised of Sequence-Segments $S$. These represent the individual Tactlines $\pi_{Tactline}(S)$ used to carry out production and are thus contiguous sections along the path traveled by a component. Within these sections the flow of materials is driven by the common production system Tact. The relative order of the Sequence-Segments is encoded by the integer $SIdx$, which corresponds to a Segment's position in the Sequence where $1 \leq SIdx \leq n$ with $n$ being the total number of Segments in the Sequence. The full set of Sequence-Segments for all Jobs are contained in the **Segment Container** $SC$. Thus the Tactsequence for a Job $j$ is $TSeq_j = \sigma_{(JID=j.JID)\wedge(TSeqID=j.TSeqID)}(SC)$.

$$S = \{JID, TSeqID, SIdx, TLID\} \qquad (5.15)$$

The constituting elements of a Sequence-Segment are Tactstation-Requirements. Each of these presents a Tactstation traversed by a Job on its way through the production facility. The length of a Sequence-Segment $|S|$ is defined as the number of contained Tactstation-Requirement $TSR$. It can either be equal to one if the Segment represents an Isolated Manufacturing Step. Alternatively it can be greater than one and simultaneously equal to $|\pi_{TLID}(S)|$ if the Segment represents a Tactline. Tactstation-Requirements are in turn comprised of zero, one or multiple alternative Tactsubstation-Requirements $TSSR$. A Tactsubstation-Requirement contains both information encoding the required Processing Capacity (RequiredCapacity) and the required Machine setup (RiggingKey). Since the latter describe the processing required per Tactstation, they directly correspond to the Process Steps from which they were derived during Tactsequence Mapping. If a Job were to pass through a Tactline consisting of three Tactstations and only requires processing on the first and last Tactstation, it would feature non-empty Tactstation-Requirements only for the first and third Tactstations.

$$TSR = \{JID, SIdx, TSID\} \tag{5.16}$$

$$TSSR = \{JID, SIdx, TSID, TSSID, RiggingKey, RequiredCapacity\} \tag{5.17}$$

Finally the complete sets of Tactstation Requirements are contained in the **Tactstation-Requirement Container** $TSRC$ while all Tactsubstation Requirements are contained in the **Tactsubstation-Requirement Container** $TSSRC$.

# Tactsequence Mapping

Tactsequence Mapping serves to generate a Sequence of Tactlines which are capable of performing the Process Steps contained in $PSS_{routing}$ for every Jobs in the Job Pool. Thus it represents a preparatory step necessary for the commencement of daily production planning activities. It builds upon completed long term production planning activities such as the design and layout of the synchronized manufacturing system (refer to Chapter 2.3.3).

In the production engineering step of the Tool Manufacturing Process Chain prior to production planning (refer to Chapter 2.4), expert knowledge is used to determine the series of process steps necessary to manufacture all individual tool components. This involves selecting the technological sequence of necessary production steps, choosing suitable production machines and deriving the necessary machine capacities per machine. This step can be performed manually by knowledgeable employees with or without the assistance of software tools. Upon completion the results are combined in a so-called **Routing**. This marks the beginning of Tactsequence Mapping activities, which transform the generated routings into suitable Tactsequences.

## Preparatory Steps

Prior to commencing with the actual generation of Tactsequences two preparatory activities must be completed. The first is the definition of the parameters, which will govern the execution of the generation of Tactsequences (refer to Chapter (refer to Chapter 5.2.1.1). This includes setting upper bounds for Tactsequence properties such as the number of Tactpallet loading operations or the prioritization of objectives for the optimized generation of Tactsequences. The second activity, the pre-processing of all input data, serves to eliminate superfluous elements from the input data and ensure its consistency (refer to Chapter (refer to Chapter 5.2.1.2).

### Parameterization

Performing a suitable parameterization ensures that the algorithmic flow and results reflect both management priorities and external restrictions. This section will present all of the available parameters and describe their respective effects on the ensuing Tactsequence Mapping.

One of the fundamental decisions is whether or not to permit the use of *Isolated Manufacturing Steps* within a generated Tactsequence. Since an Isolated Manufacturing Step has a duration of one Tact and covers exactly one processing step in the routing, its usage always represents a perfect fit. When viewing the Tactsequence Mapping problem from the perspective of throughput-time optimization, assembling a Tactsequence exclusively from Isolated Manufacturing Steps would provide the perfect result: a Tactsequence equal in length to the series of process steps contained in the routing. Due to the reduced logistics effort and increased transparency on the shop-floor associated with the use of Tactlines and Tactpallets, this solution is, however, usually not ideal from a more balanced perspective. Thus three parameters are provided to allow for fine-grained control of the use of Isolated Manufacturing.

| | | |
|---|---|---|
| $TSM_{UIP}$ | Use of Isolated Manufacturing | 0 to disable, 1 to enable |
| $TSM_{API}$ | Allow pure Isolated Manufacturing Tactsequences | 0 to disable, 1 to enable |
| $TSM_{AMI}$ | Absolute maximum of Isolated Manufacturing Steps | $\in \mathbb{N}$ |

Mapping of the Process Step Series of Job *j*

"perfect fit"                                With Idle Tacts

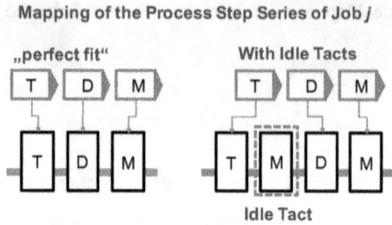

Idle Tact

Figure 5.2: Mapping with perfect fit and Idle-Tact

Another aspect requiring attention is the *Tactsequence Dilation* effect, which can be caused by Tactsequence Mapping. Finding a Sequence of Tactlines and potentially Isolated Manufacturing Steps which cover the Process Steps contained in a routing does not necessarily guarantee a perfect fit. It is entirely possible and in practice often the case that a Tactline, which is used in a Sequence-Segment, contains a number of Tactstations, which is greater than the number of covered Process Steps. On the shop-floor this leads to components traveling past certain Tactstations in a Tactline, which they do not actually require. Tacts in which a component does not receive processing on a Tactstation while moving along a Tactline are hence referred to as an *Idle-Tact*. Figure 5.2 shows an example in which Process Step Series fits the red Tactline perfectly. The blue Tactline would also be permissible but would require the Job to wait on the Tactpallet without processing for one Tact. While this effect is inherent in batching similar but non-identical Process Step Series, it does lead to increased throughput-time relative to the theoretical ideal. Thus the two parameters, absolute and relative Tactsequence dilation are provided to limit the level of permissible dissimilarity between process step series and Tactlines, which in turn leads to an dilation of the Tactsequence.

| | | |
|---|---|---|
| $TSM_{ATSD}$ | Maximum absolute Tactsequence dilation, which is equal to the total number of Idle-Tacts. | $\in \mathbb{N}$ |
| $TSM_{RTSD}$ | Maximum relative Tactsequence dilation, which is equal to the ratio of total Idle-Tacts and Tactsequence length. | $\in \mathbb{R}, >= 1.0$ |

The number of logistics operations required on the factory floor, which can be influenced during Tactsequence mapping, result directly from loading Tactpallets and transporting components to Isolated Manufacturing Steps. Thus the logistical effort to process a Tactsequence is directly proportional to the number of Sequence-Segments generated during Tactsequence Mapping. To preclude the generation of Tactsequences which would prove unfeasible to handle on the factory floor the $TSM_{MLO}$ parameter allows for the provision of an absolute cap to the number of logistic operations.

$TSM_{MLO}$    Maximum number of Tactpallet loading operations    $\in \mathbb{N}$

While all of the aforementioned parameters serve to limit the size of the solution space to search during Tactsequence Mapping, no prioritization regarding the inherent conflict of objectives has been provided yet. Since Tactsequence generation directly impacts a Job's total throughput-time and logistics effort, it is however essential to be able to compare solutions while executing the optimized Tactsequence Generation. To render prioritization decisions economical, the approach chosen in this thesis is to calculate the directly resulting costs and to choose the alternative, which minimizes overall cost. Thus a tuple *TSeqCosts* is defined, which consists of inventory costs per Tact[265] (*ICost*), the logistics cost of loading a component onto a single Tactpallet (*TPLCost*) and finally the logistics costs of transporting a component individually for Isolated Manufacturing (*IPLCost*). Due to varying value of materials, size and weight, these costs are not necessarily uniform across all Jobs. Therefore costs are provided per Job and contained in the relation $TSM_{OP}$.

With

$$TSeqCosts = \{JID, ICost, TPLCost, IPLCost\} \tag{5.18}$$

the relation $TSeqCosts$ contains costs $TSM_{OP}$ for all Jobs in $JP$

$$\tag{5.19}$$

**Input-Data Processing**

The processing of input-data serves to condition the data provided by production engineering for the ensuing Tactsequence Generation (refer to Chapter 5.2.2). The two issues dealt with during this step stem from the inclusion of Virtual and External Process Steps in the routing provided by production engineering (refer to Table 5.1).

An issue often encountered in industrial practice lies in the inclusion of Process Steps in

---

[265]In practice inventory costs per Tact result primarily from the cost of employed capital for raw materials

the routing, which do not represent a manufacturing step leading to a physical change in the semi-finished product. If, in addition to this, the component is also not required to be physically present for the completion of said step, it will henceforth be referred to as a *Virtual Process Step*. As a general rule it is not expedient to include Virtual Process Steps in the routing for assignment to a Tactsequence. This would cause these to be mapped to a Tact and cause a unnecessary extension of the throughput-time. Thus Input-Data Processing removes all Virtual Process Steps from the routing.

In this context it is however important to note that Virtual Process Steps fall into two categories: those which have no implications for the production process whatsoever and those which produce input necessary for subsequent manufacturing processes. An example for the former is the process step "goods receipt" which is often included in the routing although it does not directly influence the production process. Examples for virtual Process Steps which provide input for subsequent production steps are the "fabrication of production resources" (e.g. electrodes for erosion processes), "NC-programming of machines" or the "purchase of outsourced items". Since these steps provide essential input for successive production process steps, it is important that the necessity for their timely completion is logged appropriately. One possibility is the annotation of the impacted Non-Virtual Process Steps. This allows for the just-in-time scheduling of the eliminated Virtual Process Steps after completion of Production Scheduling.

Another problem is posed by the inclusion of *External Process Step* in the routing. Unlike Virtual Process Steps, these cause physical alteration of the semi-finished product and thus may not simply be removed from the routing. To be able to incorporate External Process Steps into a synchronized production system it is important that the steps have predefined processing and logistics time. In practice this is achieved through supplier qualification in conjunction with a contractually guaranteed service level. The Tactline layout and Machine lists should include a set of generic Tactlines and Machines for external processing. Each of these represent a possible duration for an interruption due to external processing. During Input-Data Processing External Process Steps, which contain information concerning the supplier handling the processing, are substituted with a set of process steps requiring a generic external machine, which matches the number of Tacts required for external processing.

Completion of Input-Data Processing yields the Process Step Series $PSS$, which serves as input for the subsequent Tactsequence Generation and is linked to the Job via the $PSSID$ as described on Page 68.

| Process Step Type | Description | | Action |
|---|---|---|---|
| Virtual Process Step | no physical modification product not required | no effect on subsequent process steps | removal from routing |
| | | required input for subsequent process steps | removal from routing and addition of annotation to affected process step |
| External Process Step | processing performed outside of synchronized production system | | substitution with generic process steps |

Table 5.1: Processing of Virtual and External Process Steps

## Mapping Algorithm

Tactsequence mapping serves to generate suitable Tactsequences from the routings provided by production engineering in accordance with the parameters and optimization objectives described in Chapter 5.2.1.1. This Chapter initially provides an overview of the criteria determining Tactsequence suitability, which are derived from the parameters introduced in Chapter 5.2.1.1. Subsequently it presents an algorithm which generates an optimal Tactsequence for a given routing.

### Tactsequence Suitability

In order for a Tactsequence $TSeq$ to be a valid representation of a Process Step Series $PSS$ certain conditions must be fulfilled. In general $TSeq$ is only considered to be a suitable representation of a $PSS$, if its Sequence-Segments contain compatible Tactstation- und Tactsubstation-Requirements for each contained process step $PS$. Given a Job $j$ and its associated Process Step Series $PSS_j$ the Tactsequence is defined as:

$$TSeq := (TSeq \bowtie TSRC \bowtie TSSRC \bowtie TSC \bowtie TSSC).\qquad(5.20)$$

$TSeq$ is considered a valid representation of $PSS$ if and only if

$$\emptyset = \psi_{(MID,MID)}^{(SIdx,MID),(SIdx,TSID,MID)}(PSS,TSeq).\qquad(5.21)$$

To ensure the basic suitability of a Tactsequence, its satisfaction of the parameters specified in Chapter 5.2.1.1 must now be checked. To this end two mutually exclusive preliminary checks are performed. If Isolated Manufacturing is not permitted ($TSM_{UIP} == 0$) then all Sequence-Segments must have more than one Tactstation-Requirement.

$$\forall t \in \gamma_{SIdx,Count(SIdx)}(\delta(\pi_{SIdx,TSID}(TSeq_r)))$$
$$\pi_{Count(SIdx)}(t) > 1 \qquad\qquad (5.22)$$

If Isolated Manufacturing is permitted, and pure Isolated Manufacturing Tactsequences are not permitted ($TSM_{API} == 0$) at least one Tactsequence-Segment consisting of more than one Tactstation-Requirement must exist.

$$\exists t \in \gamma_{SIdx,Count(SIdx)}(\delta(\pi_{SIdx,TSID}(TSeq_r)))$$
$$\pi_{Count(SIdx)}(t) > 1 \qquad\qquad (5.23)$$

Should both preconditions based on the chosen parameters be satisfied, the final step to ensuring suitability is checking compliance with the remaining relative and absolute limits specified during Parameterization (refer to Chapter 5.2.1.1). To this end four helper functions $\phi_{relSD}$, $\phi_{absSD}$, $\phi_{IPS}$ and $\phi_{LO}$ are defined, which extract the relative sequence dilation, absolute sequence dilation, number of Isolated Manufacturing Steps and number of tactpallet loading operations respectively. Using these, a Tactsequence is considered suitable if and only if

$$\phi_{relSD}(TSeqID) \leq TSM_{RTSD} \qquad\qquad (5.24)$$
$$\phi_{absSD}(TSeqID) \leq TSM_{ATSD} \qquad\qquad (5.25)$$
$$\phi_{IPS}(TSeqID) \leq TSM_{AMI} \qquad\qquad (5.26)$$

and

$$\phi_{LO}(TSeqID) \leq TSM_{MLO} \qquad\qquad (5.27)$$

apply. Thus if a Tactsequence reflects the series of process steps and conforms to all constraints stemming from Parameterization it is considered to be a suitable representation of the process steps contained in a Job's routing information.

## Generation Algorithm

After describing the necessary preparatory steps and the constraints of Tactsequence Mapping in Chapters 5.2.1 and 5.2.2.1, this section now presents an algorithm developed for the generation of the optimal Tactsequence for a given routing.

Given Process Step Series $PSS$ containing $n$ Process Steps $P_i$ with $1 \leq i \leq n$, each Step can be represented by either a Tactline or an Isolated Manufacturing Step. Since Input Data Processing has been completed, all of the steps now remaining in the series are essential to the successful completion of the manufacturing process and require the physical presence of the semi-finished product. Thus each step must be represented by one of the Sequence-Segments $S_i$, which comprise a Tactsequence. In principle each Process Step can be represented by any Tactline which contains at least one Tactstation, which in turn contains at least one Tactsubstation capable of performing the necessary processing operations.

A Process Step $PS$ is represented by a $TL \in TLC$ if and only if

$$\exists TS \in TL \tag{5.28}$$

where

$$\pi_{Machine}(TS) \cap \pi_{Machine}(P) \neq \emptyset. \tag{5.29}$$

$$\tag{5.30}$$

To make a preliminary assessment of the Tactsequence generation problem's complexity, the function $repcount(x)$ is defined. This function extracts the number of Tactlines in the Tactline Container, which possibly represent a Process Step. The total number of possible Tactline combinations, which can be used to represent an entire series of process steps in a routing can thus be calculated as

$$\prod_{i=1}^{n} repcount(P_i) \tag{5.31}$$

where $n = |PSS|$.

Thus for a Process Step Series, for which a suitable representation exists, the number of possible solutions can easily grow exponentially with the increasing Process Step Series length. Since in industrial practice a Machine is generally part of multiple Tactlines, this is almost always the case.

In the estimation outlined above every process step is represented by a full pass through a Tactline and thus receives a dedicated Sequence-Segment. On the shop-floor this would result in components being loaded onto Tactpallets and passing through an entire Tactline, only to complete a single processing step. This would result in an increase of cycle time by a factor of $\frac{|TL|}{1}$ for each processing step. Thus all Tactlines, which are not Isolated Manufacturing Tactlines (refer to Chapter 5.1), would cause massive slowdowns and increases of inventory. Aside from being completely impractical this would not be consistent with the idea of reducing logistics effort by transporting components with similar processing step sequences along a Tactline using a Tactpallet.

Hence generating a suitable Tactsequence requires more than substituting single process steps for single Sequence-Segments to satisfy Equations 5.28 and 5.29. To make efficient use of Tactlines with more than one Tactstation, the search must be expanded to include the possibility of representing contiguous ordered process step subsets $PSS_{Sub}$ with single Tactlines. Expanding the search to include every possible contiguous subset when searching of representative Tactlines further increases the already potentially exponential number of possible solutions.

A suitable approach is required to reduce the computational complexity of calculating the optimal solution to the problem described above. Since the problem falls into the category of discrete linear optimization problems, ideas for the algorithm outlined in this thesis were derived from the extensive preexisting work present in this field. More specifically the Branch-and-Bound meta-algorithm (refer to Chapter 4.1.2.1) was used as the basic framework to systematically search the available solution space. Building on this framework, a suitable pruning system based on the principle of Pareto Optimality, which yields significant pruning of the solution space as the algorithm progresses, was developed.

As explained in Chapter 4.1.2.1 the general underlying idea is to systematically break down a complex problem into less complex subproblems recursively (Branch). Prior to branching again, the current tree is searched for nodes, which cannot possibly yield a better solution than the best known valid solution. These can then be eliminated prior to being expanded (Bound). The three macroscopic steps of **Branch-Selection**, **Branching**, **Bound**, which are repeatedly executed until the algorithm terminates, are now described in detail in the following sections. Since the algorithm executions for multiple Jobs are independent of one another, parallelization of the workload is trivial and thus allows for efficient usage of contemporary parallel computing architectures.

| Problem Stae | Branch Storage |
|---|---|
| currently receiving processing | $B_{active}$ |
| completed processing yielding a solution | $B_{completed}$ |
| pruned during the Bound step | $B_{inactive}$ |
| processed without generating a solution | $B_{inactive}$ |

Table 5.2: Storage of Branches depending upon their states

Before delving into the three primary steps, the problem instance $A$ for the Branch-and-Bound problem must be defined. Each problem $A$ can either be broken down into simpler subproblems $\{A', A'', ...\}$ or solved directly. Conceptually $A$ represents the task of matching a Process Step Series to a suitable Tactsequence in its various stages of completion. A problem contains information pertaining to work already completed and the supply of currently pending work. More specifically pending work is the subset of remaining unprocessed process steps while completed work is comprised of generated Tactsequence-Segments. Since multiple problem instances exist during the execution of a Branch-and-Bound algorithm, multiple potential "candidate-PSS" and "candidate-TSeqs" can exist in $PSC$ and $SC$ for any given Job (refer to Page 68). The specific instances, henceforth referred to as $A_{PSS}$ and $A_{TSeq}$, are linked to a problem via $PSSID$ and $TSeqID$.

$$A = \{AID, JID, PSSID, TSeqID\} \tag{5.32}$$

For a given Job $j$ the root of the problem tree is created from the Process Step Series $\sigma_{(j.JID=JID)\wedge(j.PSSID=PSSID)}(PSC)$, which was generated during Input Data Processing (refer to Chapter 5.2.1.2). The initial Tactsequence at the tree root is empty. In general, problems or branches can exist in one of four states, which are reflected in the problem's or branch's storage. Table 5.2 shows an overview of both. The assignment of branches to containers is elaborated upon in the algorithm description.

**Branch Selection**

The Branch Selection step serves to choose the next problem $A$ in $B_{active}$[266] to be processed. This chosen problem is either solved or broken down into a set of simpler subproblems.

---

[266]The terms "Problem" and "Branch" are used interchangeably since all Problems in $B_{active}$,$B_{inactive}$ and $B_{complete}$ are Branches in the problem tree

When it comes to selecting a suitable problem - or rather a branch in the problem tree - literature review in the area of operations research shows that little is known about good selection rule[267]. Relevant work in this area focuses on probabilistic subproblem selection[268], on the definition of subintervals, which are expected to lie in close proximity to the optimal solution[269]. Computer science literature on the other hand presents a wide array of selection or search generic strategies. Among these are both simple strategies such as depth, breadth and lowest cost first search as well as more sophisticated approaches such as $A^*$ search or iterative deepening[270].

The sequence generation problem and the associated computational complexity can be reduced significantly by pruning comparable problems in the *Bound* step. Therefore the selection strategy employed in the context of this thesis maximizes comparability. Since none of the aforementioned work focuses on improving "prunability" describing these in depth is not within the scope of this thesis. As will be shown in the Bound Section, a comparison of two subproblems becomes possible, if the remaining number of Process Steps to be processed is equal. Thus to favor comparability the subproblem, which has the largest number of non-processed[271] Process Steps, is chosen.

$$B_{wCount} = \gamma_{AID,Count(AID)}(B_{active} \bowtie PSC) \tag{5.33}$$

$$B_{maxCount} = max_{Count(AID)}(B_{wCount}) \tag{5.34}$$

Choose arbitrary tuple from

$$B_{active} \bowtie B_{maxCount} \tag{5.35}$$

**Branch**

After selecting a problem $A$ from $B_{active}$ during Branch Selection, the Branch operation breaks it down into a set of subproblems $\{A', A'', ...\}$, which are easier to solve. At the conclusion of this step the subproblems are added either to the set of active branches $B_{active}$ or inactive branches $B_{complete}$. The former is the case if a problem requires additional branching. Should the step have yielded a potential solution the child is added to $B_{complete}$.

---

[267]Cf. DÜR AND STIX (2005) Journal of Computational and Applied Mathematics, p. 67.
[268]Cf. Ibid..
[269]Cf. CSENDES (2001) Journal of Global Optimization, p. 307.
[270]Cf. POOLE AND MACKWORTH (2010) Artificial Intelligence: Foundations of Computational Agents, pp. 79; ZHANG (2012) State-Space Search: Algorithms, Complexity, Extensions, and Applications, pp. 41.
[271]Non-processed in this context refers to those Process Steps in the Process Step Series, which have not yet been mapped to a Tactline and thus have still require assignment

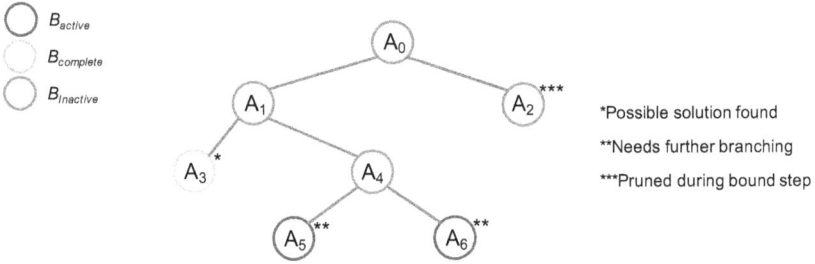

Figure 5.3: Example-State during Tactsequence Mapping

The original parent-problem $A$ is always moved from $B_{active}$ to $B_{inactive}$ irrespective of the child-problems' fate(refer to Figure 5.3). Since for a given problem $A$ this step actually performs the aforementioned conversion it represents the core of the Tactsequence generation algorithm.

Ideally Sequence-Segments are comprised of one or more non-empty Tactstation-Requirement $TSR$. The generation of Tactsequences must at least attempt to match a subset of one or more process steps of $A_{PSS}$ to a Tactline. Should this matching attempt be successful, a suitable ordered set of Tact-Requirements is assembled yielding a Sequence-Segment $S$. This matching must be performed for all Tactlines defined in the Tactline-Container $TLC$, to properly cover the solution space provided by the synchronized production system.

Thus all possible subsets of $A_{PSS}$ are generated and must be compared with all Tactlines in the Tactline-Container $TLC$. By calculating the Cartesian product of the subsets and the Tactlines, a set of branch-candidate tuples is created and placed in a set defined as $TLLN$. Branches or new subproblems are then finally created and added to $B_{active}$ for all branch candidates, where the contained process step subset $PSS_{subset}$ is compatible with the contained Tactline $TL_{match}$. Compatibility, henceforth denoted by the symbol $\leftrightarrows$, is elaborated upon in the following paragraph.

With $A$ as the problem selected for branching

$$MaxPS = count(A \bowtie PSC) \tag{5.36}$$

$$TLL = \gamma_{TLID, count(TLID)}(\sigma_{TSID \leq MaxPS}(TLC \bowtie TSC)) \tag{5.37}$$

$$MaxTLL = max(\pi_{count(TLID)}(TLL)) \tag{5.38}$$

$$TSIDSeq = \rho_{MaxSIdx/sequence(MaxTLL)}sequence(MaxTLL) \tag{5.39}$$

$$TLLN = \sigma_{(count(TLID) \leq MaxSIdx)}(TLC \times TSIDSeq) \tag{5.40}$$

$\forall x \in TLLN$ extract the subset of PSS and the Tacline to compare...

$$PSS_{subset} = \sigma_{(JID=A.JID) \wedge (PSSID=A.PSSID) \wedge (SIdx <=x.MaxTSID)}(PSC) \tag{5.41}$$

$$TL_{match} = x \bowtie TLC \bowtie TSC \bowtie TSSC \tag{5.42}$$

...and finally create new branches where

$$TL_{match} \leftrightarrows PSS_{subset}. \tag{5.43}$$

Compatibility of a $PSS_{subset}$ and a Tactline $TL_{match}$ is present, if and only if $TL_{match}$ contains Tactstations with Tactsubstations capable of processing all steps in $PSS_{subset}$ in the order in which they are contained therein. It is important to note, that this does not necessarily entail that every Tactstation in $TL_{match}$ is required by $PSS_{subset}$. In this case the components are loaded onto the Tactpallet and travel along the Tactline from beginning to end, while only being processed on certain Tactsubstations. While at first sight this might appear inefficient due to the associated extension of throughput-time, the advantage of reduced logistics effort associated with the usage of Tactpallets often outweighs this disadvantage in industrial practice. To ensure the number of Idle-Tacts or the dilation of throughput-time does not become excessive, the dilation limits introduced in Chapter 5.2.1.1 are later used in the Bound phase. In this phase, however, a Tactline is considered compatible with a $PSS_{subset}$ irrespective of the number of Idle-Tacts.

Checking for compatibility involves comparing two ordered sequences - one comprised of Tactstations, the other of Process Steps. The ordered sequences can be extracted from the non-ordered $PSS_{subset}$ and $TL_{match}$ sets by sorting these by $SIdx$ and $TS$ respectively. From its basic structure, this problem is similar to the class of sequence or string alignment problem discussed in Chapter 4.2. The specific problem requires an approach which satisfies the following requirements:

1. The approach must be tolerant to differing lengths of the compared sequences, since there is no guarantee that $PSS_{subset}$ and $TL_{match}$ are going to be of equal length.

2. The approach must take into account, that while all process steps in $PSS_{subset}$ must be processable by the $TL_{match}$ in the given order, the compatible Tactstations in $TL_{match}$ must not be a strictly contiguous sequence.

3. Since the Branch operation serves to generate Sequence-Segments, i.e. the solutions of the alignment problem, it is important not only to determine if a suitable alignment exists, but also to allow for the extraction of an optimal alignment.

Application of the first requirement eliminates the Hamming Distance since it can only be applied to binary sequences of identical length. The second requirement renders the application of the longest common substring problem impractical. Finally the longest common subsequence problem allows for verification of $PSS_{subset}$ processability by checking if $LCS(PSS_{subset}, TL_{match}) = Count(PSS_{subset})$. Common algorithms however yield the longest common subsequence and not an optimal alignment of the two (refer to Chapter 4.2.3). In addition to satisfying the first two requirements, the Levenshtein distance discussed in Chapter 4.2.4 allows for extraction of an optimal alignment via the edit transcript. It will thus serve as the foundation of $PSS_{subset}$ to $TL_{match}$ matching and alignment.

While Levenshtein's similarity metric fulfills all requirements listed above, the distance metric requires some adjustments before proceeding. In principle, the permissible operations to transform one sequence into another encompass insertions, deletions and substitutions. When comparing a $PSS_{subset}$ to a $TL_{match}$ in this context however, the objective is to determine if processing a Job is possible or can become possible by inserting Idle-Tacts. Thus the substitution and deletion operations are not permissible regarding $PSS_{subset}$. In line with the compatibility definition for Process Step Series and Tactlines, compatibility on the Process Step and Tactstation level is simply defined as:

With $PS_{relevant}$ and $TSS_{relevant}$ defined as

$$PS_{relevant} = \sigma_{(JID=p.JID),(PSSID=p.PSSID),(SIdx=p.SIdx)}(PSC) \qquad (5.44)$$

$$TSS_{relevant} = \sigma_{(JID=p.JID),(PSSID=p.PSSID),(SIdx=p.SIdx)}(PSC). \qquad (5.45)$$

compatible(p,t) is defined as follows:

with $p \in PSC$ and $t \in TSC$ these are deemed compatible if and only if

$$\emptyset \neq (PS_{relevant} \bowtie_{MID=MID} TSS_{relevant}). \qquad (5.46)$$

For the purposes of this thesis the Fisher-Wagner algorithm presented in Chapter 4.2.3 is used as a baseline. The two input strings $S_1$ and $S_2$ are substituted with the Tactline $TL_{match}$ and process step sequence $PSS_{subset}$ respectively. To ensure deletions and substitutions are never chosen as feasible options, the costs for these are increased from 1 in the standard implementation to $\infty$. Furthermore the equality operator is replaced with the aforementioned compatibility definition. A matrix $D$ with the dimensions $Count(TL \bowtie TSC) \times count(PSS_{subset})$ is defined where the top and left borders of the matrix are filled as follows

$$D(i, 0) = \infty \qquad\qquad \text{for } i \geq 1 \qquad\qquad (5.47)$$

and

$$D(0, j) = j \qquad\qquad \text{for } j \geq 0. \qquad\qquad (5.48)$$

Furthermore the rules used during algorithm execution are adjusted as follows

$$D(i, j) = min \begin{cases} D(i-1, j) + \infty \\ D(i, j-1) + 1 \\ D(i-1, j-1) + eq(i, j) \end{cases} \qquad\qquad (5.49)$$

where

$$eq(i, j) = \begin{cases} \infty : \neg compatible(first(\sigma_{(TSID=i)}(TL_{match})), first(\sigma_{(SIdx=j)}(PSS_{subset}))) \\ 0 : compatible(first(\sigma_{(TSID=i)}(TL_{match})), first(\sigma_{(SIdx=j)}(PSS_{subset}))). \end{cases}$$

$$(5.50)$$

After execution of the Fisher-Wagner algorithm modified as described above, the Levenshtein Distance of Tactline and Process Step Series $LD_{TL_{match}, PSS_{subset}}$ is contained in $D(count(PSS_{subset}), Count(TL \bowtie TSC))$, the bottom right cell of the matrix $D$. If $LD_{TL_{match}, PSS_{subset}} = \infty$ the process step series cannot be matched to the given Tactline since deletion or substitutions with transaction costs of infinity were obviously necessary to generate an alignment.

$$PSS_{subset} \leftrightarrows TL_{match} \qquad\qquad (5.51)$$

if and only if

$$LD_{TL_{match}, PSS_{subset}} < \infty \qquad\qquad (5.52)$$

If Equation (5.52) is satisfied, the Process Step Series is compatible and can be matched to the Tactline $TL_{match}$ with a number of Idle-Tacts equal to the Levenshtein Distance.

As stated in the introductory paragraph of this section, the superordinate goal of the Branch operation is the generation of Tactsequence-Segments $S$ for a Process Step Series subset $PSS_{subset}$ compatible with $TL_{match}$. When a new branch $A_1$ is created for a branch-candidate tuple $A_{parent}$, a corresponding Tactsequence-Segment is derived from an arbitrary co-optimal alignment or its corresponding edit transcript. Given compatibility, a co-optimal alignment can be extracted by applying the traceback algorithm discussed in Chapter 4.2.3 with the following modifications:

$\nwarrow$ if and only if both

$$D(i,j) = D(i-1,j-1) \tag{5.53}$$

and

$$compatible(first(\sigma_{(TSID=i)}(TL_{match})), first(\sigma_{(SIdx=j)}(PSS_{subset}))) \tag{5.54}$$

$\leftarrow$ if and only if

$$D(i,j) = D(i,j-1) + 1 \tag{5.55}$$

Traceback need only be applied to ascertain a suitable alignment when compatibility is given (as defined in Equation (5.52)). Thus Equations (4.9) and (4.10) require no modification as these only apply to substitutions or deletions. To ease the process of constructing a Tactsequence-Segment a transformation function $\zeta$ is defined. $\zeta$ transforms a process step $P$ into a Tactstation-Requirement $TSR$ containing a set of Tactsubstation-Requirements $\{TSSR_1, ...., TSSR_n\}$ for each machine capable of performing the necessary processing operation. A co-optimal alignment is produced by commencing traceback in the bottom right corner and iteratively moving to the top left by choosing the directions indicated by Equations (5.53) and (5.55). The corresponding Tactsequence-Segment can now easily be constructed by starting with an empty sequence of Tactstation-Requirements $S$ and appending an empty Tactstation-Requirement to the beginning for every leftward movement $\leftarrow$. For every diagonal movement $\nwarrow$ a Tactstation-Requirement is generated using $\zeta$ and concatenated with the beginning of $S$. Thus with $\oplus$ as the concatenation operator, $t$ as the current iteration and $D(i,j)$ as the current matrix cell, $S$ is constructed by executing $S = TSR_{|TL|-t} \oplus S$ with

$$TSR_{|TL|-t} = \begin{cases} \zeta(PSS_{subset}(PSSCnt - t - D(i,j) + D(PSSCnt, TLCnt)) & : \nwarrow \\ \emptyset & : \leftarrow \end{cases}$$

$$(5.56)$$

and

$$PSSCnt = count(PSS_{subset}) \text{ and } TLCnt = count(TL_{match})$$

with each iteration.

After generating new Branches $A_1...A_n$ for a problem $A_{parent}$, the latter is moved from $B_{active}$ to $B_{inactive}$. Furthermore the compatibility of $PSS_{subset}$ and $TL$ is evaluated as outlined above for each candidate. New Branches are created for all compatible candidates. In a new Branch, $PSS$ is equal to the remaining Process Steps not assigned in this Branch operation. $TSeq$ is equal to $\pi_{TSeq}(A_{parent}) \oplus S$ where $S$ is the corresponding new Sequence-Segment generated for $PSS_{subset}$ of $A_{parent}$. Should a successful Branch generation yield a branch $A_1$ with a non-empty series of remaining process steps ($count(\pi_{PSS}(A_1)) \geq 1$) the Branch is added to $B_{active}$. Otherwise an empty $PSS$ indicates that a possible solution for the assignment problem has been found.

**Bound**

After completing the expansion of a selected active branch in $B_{active}$ the Bound step serves to safely eliminate those target branches which cannot possibly contain an admissible solution which is better than a valid solution contained in $B_{complete}$. This step reduces both computational complexity and time by avoiding unnecessary expansion and evaluation of candidate branches. The Bound step can be broken down into two phases:

1. Elimination of non admissible branches
2. Elimination of non Pareto optimal branches

The first phase "Elimination of non admissible branches" (refer to Item 1) serves to remove all branches from $B_{active}$ and $B_{complete}$, which are not admissible in their current state and, given further expansion, can never yield an admissible solution. In Chapter 5.2.2.1 the criteria which are used to evaluate the admissibility of a Tactsequence as a representation of a Process Step Series were first introduced. Thus the current admissibility of a branch is easy to check by applying all criteria. For branches contained in $B_{complete}$ this is sufficient since these have completed processing and thus will remain invalid in future iterations if this is currently the case. For branches in the $B_{active}$ matters are more complex and only a subset of the criteria can be used to eliminate invalid branches, since determining future admissibility is impossible for some criteria prior to completion.

Concerning the restrictions placed on Isolated Manufacturing reflected in $TSM_{UIP}, TSM_{API}, TSM_{AMI}$ (refer to Equations (5.22), (5.23) and (5.26)) only the eschewal can be enforced. This is due to the fact that the number of Isolated Manufacturing Steps in $\pi_{TSeq}(A_i)$ increases monotonously during branching operations. The same reasoning applies to the limits on Tactsequence dilation and Idle-Tacts $TSM_{ATSD}, TSM_{MLO}$ (refer to Equations (5.24), (5.25) and (5.27)). Here the absolute Tactsequence dilation and number of logistics operations increase monotonously while performing the Branch-and-Bound operations. Relative Tactsequence dilation on the other hand can also decrease when an new Sequence-Segment is added. Since branches in the $B_{active}$ set must eventually be moved out of this set the algorithm structure ensures that a final Bound operation is executed after the final Branch operation. Therefore the application of all criteria is guaranteed and thus $B_{complete}$ can only contain admissible solutions. When eliminating non-admissible branches the following operations are performed for all non-pruned branches:

$\forall A \in (B_{active} \cup B_{complete})$ move $A$ to $B_{inactive}$

if

$$\phi_{absSD}(A.TSeqID)) \leq TSM_{ATSD} \qquad (5.57)$$

or

$$\phi_{IPS}(A.TSeqID) \leq TSM_{AMI} \qquad (5.58)$$

or

$$\phi_{LO}(A.TSeqID) \leq TSM_{MLO} \qquad (5.59)$$

or

$$(TSM_{UIP} == 0) \wedge (\exists t((t \in \gamma_{JID,TSeqID,count(TSeqID)}(TSeq_t \bowtie TSRC) \wedge$$
$$(t.count(TSeqID) = 1)) \qquad (5.60)$$

Additionally the following operations are executed on branches, which have completed processing:

$\forall A \in B_{complete}$ move $A$ to $B_{pruned}$

if

$$\phi_{relSD}(A.TSID) \leq TSM_{RTSD} \qquad (5.61)$$

or

$$(TSM_{UIP} == 0) \wedge (\forall t((t \in \gamma_{JID,TSeqID,count(TSID)}(TSeq_t \bowtie TSRC) \wedge (t.count(TSID) = 1)) \qquad (5.62)$$

The second phase "Elimination of non Pareto optimal branches" (refer to Item 2) serves to further reduce the number of active branches by eliminating non Pareto optimal branches. The underlying idea is to compare currently active branches with each other and to prune inherently inferior branches. Thus only the most promising branches receive processing to completion. The pruning strategy presented in this thesis achieves a reduction computational complexity without sacrificing solution optimality.

The first aspect which requires clarification in this context is comparability. Every problem instance $A_i$ or rather the contained Tactsequence $TSeq$ is characterized by the performance indicators:

1. Tactsequence dilation

2. number of Idle-Tacts

3. number of Isolated Manufacturing Steps

These again are a reflection of management priorities regarding the minimization of throughput time, Tactpallet loading operations and number of Isolated Manufacturing Steps (refer to Chapter 5.2.1.1). The challenge when comparing problem instances from $B_{active}$ lies in the fact that a problem instance, which is superior with regard to one or even all of the aforementioned properties in an iteration $iter$, is not guaranteed to spawn child-branches, which are superior to inferior problems' child branches in iteration $iter + n$. A solution is to find a scope within which comparisons can be made and only prune branches within this group. One application of this idea to the problem at hand is to limit comparisons to occur only within groups of problems, which are certain to produce identical sets of child-branches. More specifically this is the case, when problems have identical unprocessed processing steps

$\pi_{PSS}(A_i)$. Since all three performance indicators listed above (refer to Items 1 to 3) increase monotonously as processing advances an indicator-based comparison of branches within the aforementioned scope is a feasible way of identifying inferior branches.

Given

$$PSSCnts = \gamma_{AID,count(AID)}((B_{active} \cup B_{complete}) \bowtie PSC)$$    (5.63)

The set of all comparable problems is defined as

$$CompProbs = \{p \mid \exists t_1 \in PSSCnts \exists t_2 \in PSSCnts$$

$$(t_1.count(AID) = t_2.count(AID) \wedge t_1.AID \neq t_2.AID$$

$$p.AID1 = t_1.AID \wedge p.AID2 = t_2.AID\}$$    (5.64)

Aside from proper comparability, the methodology with which branches are pruned must be established. The principle used at this point is based on the concept of Pareto efficiency or optimality. The term traces back to the work of the Italian engineer and economist Vilfredo Pareto (1849-1923), who first used this concept in a micro-economic context to describe the situation of individuals in a society with maximized ophelimity. Today it is also applied in neighboring fields such as mathematics or engineering to solve practical problems in banking, agriculture, health service and manufacturing[272]. Pareto optimality describes a state where no individual in a group can improve his or her situation without negatively impacting at least one other individual in the group[273]. When a system is not yet in a Pareto optimal state at least one Pareto improvement can be executed, in which the situation of at least one individual is improved without harming any others. In this case the improved state $state_{new}$ is said to dominate the previous, non optimal state $state_{prev}$ ($state_{new} \succ state_{prev}$). When a state is reached, in which no Pareto improvements are possible, a Pareto optimal state[274] has been reached.

In multidimensional optimization, a solution is considered Pareto optimal if no feasible vector of decision variables exists, which would improve the solution in at least one dimension, while having no detrimental effect on all others. Often a problem has more than one Pareto optimal

---

[272]Cf. PARDALOS, MIGDALAS AND PITSOULIS (2008) Pareto Optimality, Game Theory and Equilibria, p. 481; MIETTINEN (2012) Nonlinear Multiobjective Optimization, p. xiv; YOSHIMURA (2010) System design optimization for product manufacturing, p. 231.
[273]Cf. LÖCHEL (2013) Mikroökonomik: Haushalte, Unternehmen, Märkte, p. 144.
[274]Cf. KOEPPEN, SCHAEFER AND ABRAHAM (2011) Intelligent Computational Optimization in Engineering: Techniques & Applications, p. 375.

solution, the full complement of which are contained in the Pareto (optimal) set[275]. Since the optimal solution vector must always belong to the Pareto set, limiting the considered solutions to this set reduces the solution space observed at any given point in time without sacrificing solution optimality[276].

As described in Chapters 5.2.1.1 and 5.2.2.1 the assignment of a Tactsequence to a Process Step Series is a multidimensional optimization problem. The Tactsequence dilation, the number of Idle-Tacts and the number of Isolated Manufacturing Steps must be optimized simultaneously. As stated above, it is in this case sufficient to consider only solutions in the Pareto set, when searching for the global optimum. Thus the principle of Pareto optimality can be used to prune branches within the aforementioned groups with identical unprocessed processing steps. Thus when eliminating non Pareto optimal branches, the following formal definition must hold.

With *CompProbs* as the set of all comparable problems prune problems where

$$\{p \mid \exists t \in CompProbs \exists k_1, k_2 \in (B_{active} \cup B_{complete})$$

$$(t.AID1 = k_1.AID \wedge t.AID2 = k_2.AID \wedge k_1 \succ k_2 \wedge p = k_1)\} \qquad (5.65)$$

**Mapping Completion**

After having executed the Sequence Generation described in the previous Chapter, a set $B_{complete}$ of suitable, Pareto optimal Tactsequences has been generated for a Job's Process Step Series. Should this set be empty, it was not possible to generate a sequence and thus the Job cannot be produced using the synchronized production system. If the set however contains elements, the problem of choosing the "best" representation of a Job in light of the specified management preferences presents itself. While in principle there are various possible approaches such as the weighting of solution dimensions using pairwise comparison[277] and later ranking the branches, the objective in the scope of this thesis is to provide a means of choosing the most economical Tactsequence. This is reflected in the definition of $TSM_{OP}$ in Chapter 5.2.1.1. which contains the costs resulting from inventory and logistics. Inventory cost $ICost$ is provided as the cost per Job per Tact, logistics costs are defined as the cost per Tactpallet loading operation $TPLCost$ and transportation cost is the resulting from the logistics effort enabling an Isolated Manufacturing operation $IPLCost$.

[275]Cf. YANG (2010) Engineering optimization: An Introduction with Metaheuristic Applications, p. 235.
[276]Cf. STATNIKOV AND MATUSOV (1995) Multicriteria optimization and engineering, p. 4.
[277]Cf. (2014)Handbuch Qualitätsmanagement, p. 516.

Thus to maximize the efficiency of processing by the synchronized production system, the branch from $B_{complete}$, which minimizes the overall inventory and logistics costs is chosen.

For a Job $j$ with a coresponding set $B_{complete}$

$$BranchesWCriteria = \pi_{AID,\phi_{absSD}(TSeqID),\phi_{IPS}(TSeqID),\phi_{LO}(TSeqID)}(B_{complete}) \quad (5.66)$$

$$BranchesWCosts = \{r \mid \forall b \in BranchesWithCriteria \exists c \in TSM_{OP}$$

$$(j.JID = b.JID \wedge b.JID = c.JID$$

$$r.AID = b.AID \wedge$$

$$r.ICost = b.absSD \cdot c.ICost \wedge$$

$$r.TPLCost = b.LO \cdot c.TPLCost \wedge$$

$$r.ITCost = b.IPS \cdot c.IPLCost)\} \quad (5.67)$$

$$BranchesWTCost = \{r \mid \forall b \in BranchesWCosts$$

$$(r.AID = b.AID \wedge$$

$$r.TotalCost = b.ICost + b.TPLCost + b.ITCost)\} \quad (5.68)$$

Now choose branch $b$ where

$$b_{optimal} = min_{TotalCost}(BranchesWTCost). \quad (5.69)$$

After choosing a branch from $B_{complete}$ as described above, the contained Tactsequence is linked to the Job in the *JobPool* by updating the linking TSeqID attribute.

For a Job $j$ with a coresponding optimal branch $b_{optimal}$

$$j.TSeqID = b.TSeqID. \quad (5.70)$$

After updating the Job tuple, Tactsequence Generation for a given Job is complete and must now be repeated by for all remaining unprocessed Jobs in the Job Pool $JP$.

# Production Scheduling

The object of Production Scheduling is to plan the manufacturing of Jobs contained in the Job Pool. This planning must be carried out taking the adherence to due-dates into account and simultaneously maximizing machine utilization while minimizing throughput-time and inventory[278].

Due to the inherent conflict scheduling while optimizing all of the above parameters is impossible[279]. For a synchronized production system the influence on throughput-time during Production Scheduling is limited to adjusting the time Jobs spend in storage between Sequence-Segments. Since the remaining throughput-time of a Job is determined by the number of Tactstations in its Tactsequence, greater influence on throughput-time can effectively be exerted during Tactsequence-Mapping or during the setup of the production system. The minimization of throughput-time will thus not be addressed directly in this context.

This leaves two parameters for optimization: **adherence to due-dates** and **maximization of machine utilization**. In line with prior work, the methodology and algorithms presented in this thesis will give adherence to due-dates priority over machine utilization by imposing it as a hard constraint. The primary reasons for this choice lie in the increased industry focus on *just-in-time* and *just-in-sequence* supply of parts[280].

This Chapter first provides a definition of the optimization problem (Chapter 5.3.1) followed by necessary complementary definitions (Chapter 5.3.2). Based hereupon a brief outline of the proposed scheduling algorithm structure is presented (Chapter 5.3.3). Finally the last two SubChapters present detailed descriptions of both the necessary preparatory steps (Chapter 5.3.4) and the three phases of algorithm execution (Chapter 5.3.5).

## Problem Definition

Production Scheduling in the synchronized production system discussed in Chapter 2.3.3 and Chapter 5.1 is equivalent to the assignment of Tactstation- and Tactsubstation-Requirements to Tacts for all available Jobs. As stated before, a Tact is the relevant time unit. A higher level of granularity when scheduling production, e.g. scheduling the exact point in time during a

[278]Cf. (2010)Operative Exzellenz im Werkzeug- und Formenbau, pp. 16.
[279]Cf. VAN DYKE PARUNAK (1991) Journal of Manufacturing Systems, p. 243.
[280]Cf. HÜTTMEIR et al. (2009) International Journal of Production Economics, p. 501; BRAKEMEIER AND JÄGER (2004) Schlanke Produktion, p. 86; ZISKOVEN (2013) Methodik zur Gestaltung und Auftragseinplanung einer getakteten Fertigung im Werkzeugbau, p. 103.

Sequence-Segments of Job *j*

| | Segment 1 | Segment 2 | Segment 3 | Segment 4 | |
|---|---|---|---|---|---|
| | 3 Tacts | 5 Tacts | 4 Tacts | 3 Tacts | Logistic-Tact |
| | | | | | Wait-Tact |

| Allocation Segment 1 | Allocation Segment 2 | Allocation Segment 3 | Allocation Segment 4 |
|---|---|---|---|
| 3 Tacts | 5 Tacts | 4 Tacts | 3 Tacts |

Tact 0                                                                                       Tact 22
                                                                                             Due date of Job *j*

| Start-Tact of Segment 1: *Tact 1* | Start-Tact of Segment 2: *Tact 6* | Start-Tact of Segment 3: *Tact 13* | Start-Tact of Segment 4: *Tact 18* |
|---|---|---|---|

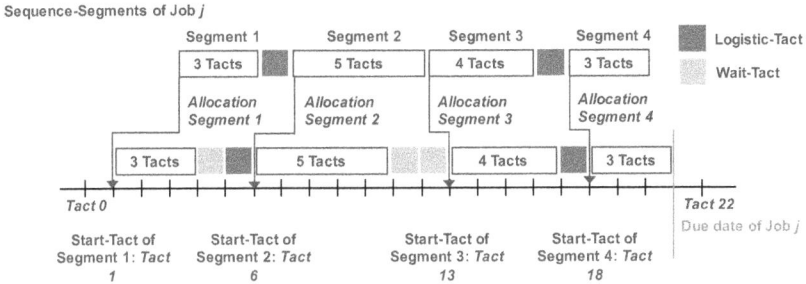

Figure 5.4: Overview of the Production Scheduling principle

Tact, would increase the complexity of the scheduling task while not improving the flow of materials on the factory floor[281].

During Tactsequence Mapping, Tactstation-Requirements are grouped into Sequence-Segments, which are contiguous sections of a Job's production process. Within these sections a Job travels along a Tactline on a Tactpallet or receives isolated processing. Since the Job must be moved to the next Tactstation within the current segment when a Tact ends, the timing of Tactstation-Requirement processing is completely determined by the Tact in which the enclosing Sequence-Segment commences processing. **Thus the central problem in Production Scheduling can be simplified to determining the Start-Tacts for all Sequence-Segments of the Jobs in the Job Pool while adhering to due-dates and maximizing machine utilization.** Figure 5.4 illustrates the assignment of Jobs' Sequence-Segments to Start-Tacts.

The problem definition contains the restriction of timely order completion and the maximization of Machine-Capacity usage as objectives. Since each Job has a due-date determined during production engineering (refer to Chapter 5.1) the latest Tact, within which production is permissible is the latest Tact on a date immediately prior to the due-date(refer to Figure 5.4). The objective is to use the available Machine-Capacity to the greatest extent possible without exceeding the total available capacity of a machine during a Tact. Thus a Sequence-Segment can only be assigned to Tact, where the available Machine-Capacities are sufficient to satisfy all requirements along the entire Segment. Making efficient use of Machine capacities entails maximizing productive Machine time and thus minimizing setup

---

[281]Furthermore this would require large amounts of additional formalized information concerning individual Jobs and the capabilities of the machines on the shop-floor. Also this would preclude workers from introducing their expertise and knowledge into the scheduling of their daily activities on a microscopic level and could easily impact worker acceptance of the production system negatively

time. In addition, Production Scheduling must also attempt to schedule Jobs with similar machine-setup requirements to arrive at Machines simultaneously.

The scheduling problem of assigning Sequence-Segments to Start-Tacts bears a similarity to the Knapsack Problem discussed in Chapter 4.3. If in a simplified version of the problem the production capacities of a single Machine in a single Tact were to be filled by assigning Jobs to it, this could be modelled using the 01-Knapsack problem. In this case the available Processing-Capacity of the Machine would be the Knapsack, while the Jobs would be the items. To replicate the scheduling problem at hand as a Knapsack Problem, the basic Binary Knapsack Problem will be gradually extended to incorporate all optimization objectives and constraints. The resulting problem is a combination of Knapsack Problem variants discussed in Chapter 4.3.2. The construction will be described in multiple steps, each of which corresponds to one of the integrated knapsack problem types.

1. **Multiple Machines and Tacts : d-Dimensional Knapsack Problem (dKP)**

   As noted in the introduction, the available production capacity on the factory floor over time represents the knapsack. Usually production facilities contain multiple machines and the regarded timespan when scheduling encompasses multiple Tacts. Since the available capacities on the various machines vary from Tact to Tact, the knapsack must feature a multi-dimensional weight constraint (refer to dKP in Chapter 4.3.2). More specifically the weight constraint is a vector consisting of one entry containing the available Machine-Capacity for every unique combination of $TID$ and $MID$ in the scenario, which is equivalent to containing one entry per tuple in the Capacity Supply $CS$.

2. **Weight dependent on Tact : Multiple Choice Knapsack Problem (MCKP)**

   The items to choose from when filling the knapsack are constructed from the Sequence-Segments in a Job's Tactsequence. Thus for each of the Segments at least one item must be generated. Given a Tactsequence for a Job consisting of $m$ Sequence-Segments, all Segments must be assigned to different Start-Tacts. Taking the Job's due-date into account, each Segment can theoretically be assigned to one of many different Start-Tacts. Since, as stated in the previous section, the required processing capacities (item-weights) depend on the chosen Start-Tact, one item must be generated for each Start-Tact candidate of every Segment in a Tactsequence. For a solution to be valid, any Sequence-Segment can however only be assigned to exactly one of the candidate Start-Tacts. To reflect this limitation, disjoined multiple-choice constraints "within a Segment" are added to the problem definition (refer to MCKP in Chapter 4.3.2).

3. **Segments interdependent : Precedence Constraint Knapsack Problem (PCKP)**
   Another complication arises from the fact that the chosen Start-Tact for a Segment can limit the remaining viable Start-Tacts for the following Segments. Thus a set of precedence constraints governing the combination of items in the knapsack must be added to the problem definition. As described in the section detailing the PCKP in Chapter 4.3.2 this can be achieved by describing the relationships in a directed acyclic graph.

4. **Rigging Optimization: Integer Knapsack Problem with Setup Weights (IKPSW)**
   An important issue affecting the effective weight or capacity consumption in the knapsack is the consideration of Machine setup times. As defined in Chapter 5.1 every Tactsubstation-Requirement features a Rigging-Key, which represents setup or rigging procedures on a Machine. Two Tactsubstation-Requirements with identical Rigging-Keys referencing the same physical Machine have identical setup procedures. Thus if two or more Jobs with identical Rigging-Keys on a specific Tactsubstation arrive at the referenced Machine for processing in the same Tact, the Machine only needs to be setup a single time to process all of these Jobs. The Integer Knapsack Problem with Setup Weights presented in Chapter 4.3.2 models this exact behavior. Regrettably however the problem description found in literature assumes, that the items receiving the setup time bonus are identical. In the scheduling case at hand, this does not apply: it is entirely possible, for two different Jobs to have identical rigging prerequisites for single manufacturing step on a Machine. Thus the setup weights cannot be defined for item-classes in this case. By defining setup-weight-vectors for each unique combination of Machine, Tact and Rigging-Key, the the IKPSW can be modified to fit the scheduling use-case.

5. **Multiple Objectives during Optimization : Multiobjective Knapsack Problem (MOKP)**
   The final aspect of the optimization problem is the maximization of machine utilization. Building on premise of Lean Manufacturing and the Theory of Constrains it is the performance of the a system's constraint which will determine the performance of the entire system[282], the optimization objective can be simplified to ensure the production system's constraints are exploited to the greatest possible extent. As shown by ZISKOVEN it is easiest to fully exploit a constraint by first attempting to assign

---

[282]Cf. WOEPPEL (2000) The manufacturer's guide to implementing the theory of constraints, p. 12.

those Jobs to it, which have the largest capacity requirement on said resource[283]. Thus a superordinate optimization would be taking the exploitation of all constraints into account. To do so, the value vector for each Job could be comprised of the required capacities on machines deemed to be constraints. Thus each vector dimension would represent the maximization objective for one of the machines constraining production.

By combining various aspects of known Knapsack Problems as described above, the scheduling problem could in theory be modeled as a **d-Dimensional Multiple Choice Precedence Constraint Multiobjective Knapsack Problem with Setup Weights**. Aside from the rather unwieldy name, the resulting problem is a combination of four strongly *NP*-complete and one weakly *NP*-complete combinatorial problems. Thus none of the solution or approximation algorithms developed for the constituting problems could be applied directly. Custom tailored algorithms would have to be developed. Due to the extreme intractability of the contained problems it is questionable, whether such development effort would yield an algorithm capable of approximating or solving the scheduling problem even in polynomial time. Furthermore assuming this were possible, it is probable that the polynomial exponents would render a problem solution impractical in light of typical problem sizes in industrial practice[284].

One of the key advantages the synchronized production system at the center of this thesis lies in its simplicity, comprehensibility and thus acceptance by stakeholders from the shop-floor up to management levels. Since the Production Scheduling algorithm is intended to support production schedulers in their daily work a comprehensible modus operandi for the algorithm represents a considerable advantage. Thus the Production Scheduling algorithm presented in this thesis builds on the excellent previous heuristic by ZISKOVEN for single Tactline scheduling and extends the scope to an entire manufacturing facility. Furthermore the presented algorithm incorporates best-practices and expert knowledge from production schedulers with extensive experience in manually scheduling Jobs for a synchronized production system manually.

---

[283]Cf. ZISKOVEN (2013) Methodik zur Gestaltung und Auftragseinplanung einer getakteten Fertigung im Werkzeugbau, p. 135.

[284]In the practical use-cases used for the purposes of validation an average planning run received the following input: $\varnothing$ number of Jobs: 302.3, $\varnothing$ number of Sequence-Segments per Job: 5.4, $\varnothing$ of candidate Start-Tacts: 13.2, $\varnothing$ of machines: 56.3, $\varnothing$ number of Tacts in scenario: 45.3. Based on the Problem Definition this input-data would yield a knapsack problem with approx. 21548 items each with a 2551-dimensional weight-vector. The number of brute-force-attempts necessary to find the optimal solution based on this data would be $(13.2^{5.4})^{302.3} = 1.77 \times 10^{1829}$

# Complementary Definitions

This Chapter introduces concepts, variables and abbreviations supplemental to those contained in Chapter 5.1.

The definition of important points in time and timespans, is shown in Figure 5.5. The point in time when a Job is admitted into the Job Pool is referred to as the **Job Admission Time** ($JAT$). Since Orders are generally acquired by the sales department continuously, the resulting Jobs are also added to the Job Pool continuously. Production Scheduling on the other hand usually takes place in regularly scheduled intervals. In industrial practice, both the daily and weekly planning rhythms are widespread. The points in time when scheduling takes place will henceforth be referred to as **Production Scheduling Time** ($PST$).

The objective of Production Scheduling for a synchronized production system is to determine, which Jobs are to commence production in a certain interval of time, called the **Scheduling Horizon** ($SH$). To allow for provident planning, i.e. the early detection of future capacity bottlenecks with sufficient lead-time, SH is often chosen to be considerably longer than the interval between scheduling runs. Given e.g. daily planning, a SH of two weeks is not uncommon. The beginning and ending of a SH, which are the first and last Tacts for which scheduling is to be performed are referred to as the **Scheduling Start Tact** ($SST$) and **Scheduling End Tact** ($SET$).

The sub-interval in SH, which covers the Tacts for which a production is to commence is referred to as the **Frozen Zone** ($FZ$). As the name implies the scheduling of those Jobs assigned to one of the Tacts in the FZ is considered immutable upon completion of a scheduling run. Intuitively the FZ usually begins with the SST and lasts up to the next PST.

Since Jobs are scheduled completely, i.e. Start-Tacts for all contained Sequence-Segments are determined, a Job scheduled to being produced on the SET will most likely require production capacities in later Tacts. Thus the final timespan considered in Production Scheduling is the **Tact Horizon** ($TH$), which contains all up to and including the last Tact, which could possibly be required by a Job in the Job Pool if it were scheduled to start production on the SET.

During Production Scheduling Jobs' Tactsequences are allocated to Start-Tacts in the Scheduling Horizon. This assignment of the Sequence-Segments contained in a Job's Tactsequence to a Tact is represented by a so-called **Allocations** ($Ac$). When a Tactsequence is first assigned to a Tact, the generated Allocation has a Status-flag of *unassigned*. As the

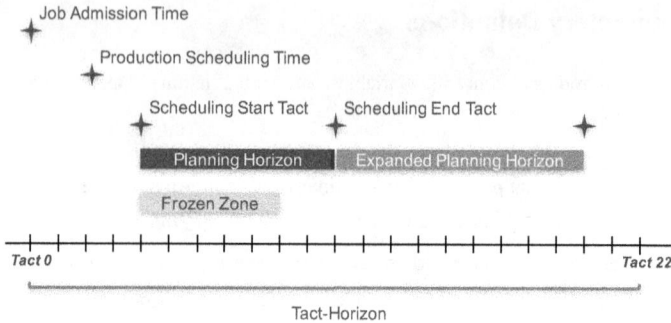

Figure 5.5: Relevant points in time and intervals during Production Scheduling

algorithm progresses, an allocation can me marked as *reversible* or *fixed*, which impacts whether the Allocation can or cannot be undone as the algorithm progresses[285]

$$Ac = \{JID, TSeqID, SIdx, TID, Status\} \tag{5.71}$$

To capture the current "state" of the scheduling algorithm, the concept of the **Scenario** (*So*), which serves as a container for all current Allocations, is introduced.

As outlined in Chapter 5.1 Sequence-Segments represent contiguous sections along the path traveled by a Job. If a Segment consists of more than one Tactstation-Requirement, it represents the traversal of a Tactline. Otherwise it represents an Isolated Manufacturing Step. Since a logistical loading operation is required upon the completion of processing for a Segment, this represents a possible point at which a Job can be transferred into temporary storage without incurring considerable additional logistics overhead. Should this be the case, the time spent in storage between the Segments of a Tactsequence is referred to as one of multiple **Wait-Tacts**. While Wait-Tacts interrupt the production flow, cause inventory buildup and thus violate the principles of Lean Manufacturing they can however have practical advantages. The idea underlying the synchronized production of individualized products is to combine Jobs on Tactpallets to simultaneously realize a high machine utilization and a constant flow of materials. This can only be achieved if the Jobs combined on a Tactpallet complement each other's capacity requirements along the entire traversed Tactline. Thus Wait-Tacts give an algorithm a little "wiggle-room" in which to shift Jobs back slightly to achieve a better alignment to other Jobs.

---

[285]In principle this classification is similar to ZISKOVEN's concept of movable components on Tactpallets (cf. Ibid., p. 132).

If a Job completes processing in the context of a Sequence-Segment, which is not the last Segment in the Tactsequence, a logistics operation is necessary to transfer it either to the next Tactpallet or to an Isolated Manufacturing Step. Since the logistical feat of completing this task between Tacts is not necessarily possible in every production setup, fixed buffer Tacts hence referred to as *Logistics-Tacts* must be factored into the production schedule.

When assigning the Segments in a Tactsequence to Tacts both Logistics- and Wait-Tacts can be inserted between Segments as described above. This can serve to optimize machine utilization or simply accommodate the existing logistics-processes in a company. These points between Segments are henceforth referred to as *Splice-Points* $(SP)$ within a Tactsequence. Splice-Points are simple tuples, which reference the Segment after which they appear (SIdx) and the total number of inserted Tacts (Count)[286].

$$SP = \{SIdx, Count\} \tag{5.72}$$

## Algorithm Outline

The algorithm presented in this thesis aims to schedule the production of Jobs for an entire manufacturing facility containing multiple, intersecting Tactlines. The heuristic algorithm outlined builds upon the previous work by ZISKOVEN which proved to deliver excellent results in real world industrial application. Thus a similar macroscopic structure was adopted when designing an algorithm capable of handling Sequences consisting of multiple intersecting Tactlines with interspersed Isolated Manufacturing Steps. The algorithm consists of three major phases, each of which will be briefly outlined in this Chapter and described in detail in Chapters 5.3.4 and 5.3.5

The first phase **Critical Job Allocation** aims to schedule time-critical Jobs for these are completed in accordance with their due-dates. Time-critical in this context refers to all Jobs in the Job Pool, for which it is necessary to commence production in the selected Scheduling Horizon to complete them in time. The algorithm presented in this thesis processes Jobs from all Tactlines simultaneously and schedules Start-Tacts for all Segments in a Tactsequence in one single iteration. While the primary objective in this phase is to find an appropriate allocation for all time-critical Jobs, the algorithm will attempt to assign all of these without violating the Target-Capacities constraint provided for all Machines.

---

[286]Count is in this case the sum of Logistics- and Wait-Tacts.

The second phase **Rigging Optimization** focuses on the rearrangement of Jobs allocated in the previous phase in order to consolidate the processing of Jobs with identical machine setup requirements. When generating Jobs, Production Engineering can optionally provide a so-called Rigging-Key for individual Tactsubstation-Requirements, which encodes the machine setup requirements for the given Job in the specified Process Step. Thus the algorithm can attempt to maximize the number of Jobs with identical Rigging-Keys processed on a Machine in a given Tact for the entire Machine Complement on the factory floor. This then simultaneously minimizes the number machine setups in a given Tact and maximizes the productive processing time.

The third and final phase **Topping Off** serves to further improve the scheduling results by allocating additional non time-critical Jobs to fill potentially available processing capacities. This is done by first choosing a set of additional Jobs, which must commence production in an expanded Scheduling Horizon. Since the objective in this phase is the maximization of machine utilization, the Jobs from the expanded SH can be chosen, which represent the best possible fit. The allocation is then chosen based on Jobs already allocated in the Scheduling Horizon and their overall Capacity-Requirement fit.

## Preparatory Steps

The three step algorithm is preceded by a series of preparatory steps outlined in this Chapter. These on the one hand involve defining all relevant parameters for the scheduling algorithm and performing preliminary tasks such as the calculation of latest possible Start-Tacts for each Job and identification of the production system's constraints on the other.

### Parameterization

The presented algorithm for Production Scheduling contains a set of parameters, which control the algorithmic flow. This Chapter provides an overview of all relevant parameters and their respective effects on the ensuing Production Scheduling.

Two basic parameters, which define the period of time for which the scheduling is to occur are the **Scheduling Start Tact** ($SST$) and the length of the **Scheduling Horizon** ($SH$). Since both of these were already defined above, please refer to Chapter 5.3.2 for a detailed description.

$PS_{SST}$    TactID of Scheduling Start Tact                          $\in \pi_{TID}(TC)$

$PS_{SH}$    Length of Scheduling Horizon in number of Tacts    $\in \mathbb{N}$

The parameter **Maximum Number of Wait-Tacts** limits the total number of Wait-Tacts (refer to Chapter 5.3.2) which can be inserted into a Tactsequence. Completely eliminating Wait-Tacts, essentially a "lean" solution, would force the scheduling algorithm to align Jobs to complement each other's capacity requirements along the entire Tactsequence consisting of multiple Segments. Inherently this is more difficult and in real-world cases yields much poorer results in terms of total system throughput. Thus allowing for Wait-Tacts adds flexibility and improves the utilization of Machine capacity. In practice a limit of two or three Wait-Tacts per Tactsequence has proven beneficial while causing only a slight extension of Jobs' throughput-time.

$PS_{MNWT}$    Maximum number of total Wait-Tacts per Tactsequence    $\in \mathbb{N}$

In some production setups the logistics effort required to transfer Jobs from one Tactpallet to the next or to deliver a Job to an Isolated Manufacturing Step cannot be handled between two production Tacts. To take this into account during Production Scheduling the **Logistics Tacts Before Entering Tactline** and **Logistics Tacts Before Isolated Processing** are provided as static parameters.

$PS_{LTBT}$    Number of Logistic-Tacts per Tactpallet loading operation    $\in \mathbb{N}$

$PS_{LTBI}$    Number of Logistic-Tacts for isolated processing logistics    $\in \mathbb{N}$

In line with the basic tenets of Lean Manufacturing and the objective of maximizing production system throughput, the algorithm seeks to maximize the exploitation of the system's constraints[287]. To do so one of the first steps during Production Scheduling is to ascertain the Machines constraining the production system's output. The **Constraint-Factor** is used in this process to determine the critical relative machine utilization above which a Machine is considered to be a system constraint (for detailed information on the procedure, please refer to Chapter 5.3.4.3).

---

[287]Cf. REVELLE (2001) Manufacturing handbook of best practices: An innovation, productivity, and quality focus / edited by Jack B. ReVelle, pp. 389; WANG (2011) Lean manufacturing: Business bottom-line based, pp. 168.

$PS_{CF}$     Constraint-Factor used to determine system constraints   $\in \mathbb{R}^{\geq 0}$

The final phase of the presented algorithm is designed to fill potentially remaining processing capacities after all time critical Jobs have been successfully allocated (Topping Off). In this scenario it would not be advantageous to select Jobs with arbitrary due dates as this could result in performing manufacturing for orders in the distant future, which would cause the build-up of ready-to-ship inventory. Thus the **Scheduling Horizon Expansion Factor** is used to limit the selection to those Jobs which require the commencement of processing in a Scheduling Horizon expanded by the Factor.

$PS_{SHEF}$     Scheduling Horizon Expansion-Factor for Topping Off   $\in \mathbb{R}^{\geq 1}$

While the number of parameters introduced in this Chapter is relatively large, most of these are determined on a long-term basis. In practice the workers responsible for day-to-day Production Scheduling must only adjust Scheduling Start Tact $PS_{SST}$, which renders the parameterization of regular production planning a manageable task.

**Calculation of Latest Possible Start-Tacts**

From an economical perspective the premature production of products leads to the build-up of inventory and is therefore not desirable. This is reflected by the industrial focus on "just-in-time" or "just-in-sequence" production. When manufacturing products with a synchronized production system, both the processing- and the throughput-time of a Job are known at the time of scheduling. Thus it is easy to calculate the latest possible Start-Tact for a Job. Since calculating this information prior to commencing scheduling simplifies the algorithm's description in Chapters 5.3.5.2 to 5.3.5.4, the calculation is presented as a preliminary step.

For each Job the due-date provided by Production Engineering is defined as the date by which the manufacturing process must be complete. Therefore the latest Tact on which production is allowed to occur can be calculated by finding the latest Tact prior to the due-date.

For a given Job $J$ the TactID of the latest production Tact is

$$LatestProductionTID = \pi_{TID}(max_{TID}(\sigma_{PIT<J.DueDate}(Cal)))\qquad(5.73)$$

The shortest possible throughput-time for a Job is given by the number of Tactstation-Requirements in the Job's Tactsequence plus the number of Logistics-Tacts required to allow for Tactpallet loading and distribution to Isolated Manufacturing Steps.

For a given Job $j$ the Tactstation-Requirement-Counts are

$$TRCnts = \rho_{count(SIdx)/Cnt}(\gamma_{JID,SIdx,count(SIdx)}((\sigma_{JID=j.JID}(JP) \bowtie TSRC))). \quad (5.74)$$

The number of processing Tacts is thus

$$PTCnt = \gamma_{JID,sum(Cnt)}(TRCnts) \quad (5.75)$$

while the number of Logistics-Tacts for Tactpallet loading is

$$LTTP = \gamma_{JID,count(JID)}(\sigma_{Cnt>1}(TRCnts)) \cdot PS_{LTBT} \quad (5.76)$$

and the number of Logistics-Tacts for Isolated Manufacturing is

$$LTTI = \gamma_{JID,count(JID)}(\sigma_{Cnt=1}(TRCnts)) \cdot PS_{LTBI}. \quad (5.77)$$

Finally for a Job $J$ the minimum throughput-time measured in Tacts is

$$TPT = PTCnt + LTTP + LTTI. \quad (5.78)$$

After determination of the Throughput-Time ($TPT$) in Tacts, the latest possible Start-Tact ($LPST$), in which the production of a Job can possibly be commenced while adhering to the given due-date, can be calculated as

$$LPST = LatestProductionTID - TPT + 1. \quad (5.79)$$

The procedure outlined is performed for all Jobs contained in the Job Pool and is stored in a **Latest Start-Tact Container** ($LSTC$) containing **Latest Start-Tact** tuples. These combine the Job-Identifier (JID) with the Tact-Identifier of the latest possible Start-Tact and a Job's Throughput-Time measured in Tacts.

$$LST = \{JID, LPST, TPT\} \quad (5.80)$$

## Constraint Identification

One of the superordinate objectives when performing the Job Allocation is to ensure that the manufacturing system's constraints are exploited to the greatest possible extent. To this end one of the preparatory steps during algorithm execution is the Constraint Identification. Due to the great importance of correctly identifying the constraints depending on the workload to be scheduled, the approach of estimating the constrains was chosen rather than requiring the constraints to be submitted as part of the parameterization.

The first step of the estimation procedure is the calculation of the capacity required by the time-critical Jobs in the Job-Pool. The Latest Possible Start Tacts calculated in the previous preparatory step (refer to Chapter 5.3.4.2) are used to isolate the Jobs $CritJobs$ which must commence manufacturing in the current Scheduling Horizon. In a second step the cumulative capacity requirements $CuCapReq$ for all Machines are calculated for all critical Jobs.

$$CritJobs = \sigma_{(PS_{SST} \leq LPST < (PS_{SST} + PS_{SH}))}(JP \bowtie LSTC) \tag{5.81}$$

$$CuCapReq = \gamma_{MID, sum(RequiredCapacity)}(CS) \tag{5.82}$$

The third step is the calculation of the cumulative capacity $CuCapSup$ offered on every Machine in the Capacity Supply in the Scheduling Horizon[288].

$$CuCapSup = \gamma_{MID, sum(TargetCapacity)}(\sigma(PS_{SST} \leq TID < (PS_{SST} + PS_{SH}))(CS)). \tag{5.83}$$

In the fourth step a load factor defined as the ratio of required capacity to available capacity is calculated for each Machine on the shop-floor.

$$LoadFactors = \rho_{\frac{CuCapReq}{CuCapSup}/LF}(\pi_{MID, \frac{CuCapReq}{CuCapSup}}(CuCapReq \bowtie CuCapSup)) \tag{5.84}$$

Given the load factors for all machines on the factory-floor, the maximum load over all Machines is ascertained. All Machines are then considered to by constraints if these have load factors, which are at least as large as the maximum load factor multiplied by the Constraint-Factor $PS_{CF}$ (refer to Chapter 5.3.4.1). The resulting MIDs of the constraining Machines are then stored in the **Constraint Container** ($CC$) for use during algorithm execution.

---

[288]While using the calculating the capacity available in the Scheduling Horizon is not a measure of the capacities the allotted Jobs will occupying precisely, this simple approximation yields good results if the available machine capacities remain relatively constant over time.

$$MaxLF = \pi_{LF}(max_{LF}(LoadFactors)) \tag{5.85}$$

$$CC = \pi_{MID}(\sigma_{LF \geq MaxLF * PS_{CF}}(LoadFactors)) \tag{5.86}$$

## Scheduling Algorithm

After having outlined the basic algorithmic structure in Chapter 5.3.3 and provided a detailed description of necessary preparatory steps in Chapter 5.3.4 this Chapter presents the Production Scheduling algorithm in detail.

### Auxiliary Algorithm Fragments

This Chapter will introduce several algorithm fragments, which represent basic building blocks of the presented algorithm and be reused frequently throughout its three phases. Please note that variables defined in the sections of this Chapter are valid only within their respective Section unless Equations are explicitly referenced.

#### Check Feasibility of Scenario

The first algorithm fragment presented checks whether the Allocations of a Scenario (either in a *fixed* or *reversible* state) can be processed with the available Processing Capacities (refer to Chapter 5.1). This check will henceforth be referred to as checking the feasibility of a Scenario ($SO$) and is divided into the steps described below. Feasibility checks are always execute a relative to a specified capacity limit ($CapacityLimit$)[289].

As the Processing Capacities for all Machines are contained in the Capacity Supply ($CS$), the first step is the extraction of the capacity requirements ($CSRMT_{interim}$):

$$CRPMT_{interim} = SO \bowtie TSR \bowtie TSSR. \tag{5.87}$$

---

[289]$CapacityLimit \in \{TargetCapacity, MaximumCapacity\}$

Since Allocations align a Segment to a Start-Tact-ID ($TID$) at the Segment-level (refer to Equation (5.71)) this alignment must be applied to the individual Tactstation-Requirements in $CSRMT_{interim}$ by generating a corrected $TID$ for each one. The corrected $TID$ can be trivially derived by adding the Segment $TID$ in the Allocation and the $TSID$ in the Tactstation Requiremens and finally substracting 1. Thus $CRPMT_{corrected}$ is defined as

$$CRPMT_{interim2} = \pi_{MID,TID+TSID-1,RequiredCapacity}(CRPMT_{interim}) \tag{5.88}$$

$$CRPMT_{corrected} = \rho_{TID+TSID-1/TID}(CRPMT_{interim2})) \tag{5.89}$$

Now the required Processing Capacity used by allocated Jobs can be aggregated to $CCRPMT$.

$$CCRPMT = \gamma_{MID,TID,sum(RequiredCapacity)}(CRPMT_{corrected}) \tag{5.90}$$

Nowt the available Processing Capacity for each Machine and Tact can be easily calculated

$$Combo_{CSCR} = CCSPM \underset{CCSPM.MID=CCRPM.MID}{\bowtie} CCRPM \tag{5.91}$$

$$AvCap_{interim} = \pi_{MID,TID,sum(CapacityLimit)-sum(RequiredCapacity)}(Combo_{CSCR}) \tag{5.92}$$

$$AvCap = \rho_{sum(CapacityLimit)-sum(RequiredCapacity)/AvailableCapacity}(AC_{interim}) \tag{5.93}$$

Finally a check is performed to identify those Tacts where the required Processing Capacity exceeds the supply.

$$MissingCapacity = \sigma_{AvailableCapacity<0}(AvCap) \tag{5.94}$$

If the relation $MissingCapacity$ is non-empty, the Scenario submitted for checking is not considered feasible. In summary a function $feasibleScenario(...)$ which features the functionality described receives a Scenario ($SO$) as input and returns the boolean ($F$), is defined as:

$$F = feasibleScenario(SO, CLimit) \tag{5.95}$$

**Check Feasibility of Allocations**

When one or multiple Allocations are generated as described in the next section, these are not added to a Scenario immediately. This algorithm fragment checks whether a set of Allocations ($ACS$) can be added to a Scenario ($SO$) safely without sacrificing the Scenario's feasibility relative to a specified capacity limit ($CapacityLimit$)[290]. The necessary steps are described below:

As a first step $ACS$ and $SO$, both relations containing Allocations are merged:

$$CheckSO = SO \cup ACS. \tag{5.96}$$

Making use of the $feasibleScenario(...)$ function defined above, the feasibility of the merged Scenario is now evaluated.

$$F = feasibleScenario(CheckSO, CLimit) \tag{5.97}$$

Using the procedure described above, a function $feasibleScAcMerge(...)$ which receives the Scenario ($SO$) and set of Allocations ($ACS$) as input is defined as:

$$F = feasibleScAcMerge(SO, ACS, CapacityLimit) \tag{5.98}$$

---

[290]$CapacityLimit \in \{TargetCapacity, MaximumCapacity\}$

**Generate Allocation for Job**

The purpose of the algorithm fragment presented in this Chapter is the scheduling of Jobs' manufacturing activities represented by a Tactsequence. More specifically this is equivalent to assigning each Segment in the Tactsequence to a Start-Tact. Information regarding this assignment is stored in a so-called Allocation (refer to Figure 5.4 and Chapter 5.3.2). As described in Chapter 5.3.2 the Logistics- and/or Wait-Tacts can be inserted into Splice-Points during allocation. In practice this results in gaps in the assigned series of Tacts.

Given a Job ($J$), a set of Splice-Points ($SPS$) and a corresponding Start-Tact-ID ($STID$) this algorithm fragment generates Allocations for the Segments contained in the Job's Tactsequence. Thus for a Tactsequence with $n$ Chapterents this fragment generates $n$ corresponding Allocations. For each $S$ in $TSeq$ an Allocation ($Ac$) is generated as follows:

First, all Tactstation-Requirements for all Segments preceding $S$ are isolated and counted.

$$SPreCnt = \gamma_{JID,TSeqID,SIdx,count(SIdx)}(\sigma_{SIdx<S.SIdx}(\pi_{JID,TSeqID}(J \bowtie SC)) \bowtie TSRC))$$
(5.99)

In a second step the number of Production-, Logistics- and Wait-Tacts occurring prior the the current Sequence are calculated

$$ProTaCnt = sum(SPreCnt)$$
(5.100)

$$LoTaTlCnt = count(\sigma_{count(SIdx)>1}) \cdot LTBT$$
(5.101)

$$LoTaIsoCnt = count(\sigma_{count(SIdx)=1}) \cdot LTBI$$
(5.102)

$$WaTaCnt = count(\sigma_{SIdx<S.SIdx}(SPS))$$
(5.103)

The resulting Start-Tact-Id $TID$ for the Segment ($S$) can thus be calculated as:

$$TID = STID + ProTaCnt + LoTaTlCnt + LoTaIsoCnt + WaTaCnt.$$
(5.104)

After calculation of $TID$ an Allocation with a status flag set to *unassigned* is generated receiving its $JID$, $TSeqID$, $SIdx$, $TID$ from the "parent-Segment". By iterative repeating the procedure described above for each Segment in the original Tactsequence, a full set of the resulting Allocations ($ACS$) can is assembled. Thus a function $genAllocation(...)$ is defined as follows.

$$ACS = genAllocations(J, SPS, STID)$$
(5.105)

**Attempt To Find Feasible Allocations**

The algorithm fragments defined above can generate Allocations for a Job and check the feasibility of assigning Allocations to a Scenario. This Chapter will leverage these two fragments to find a feasible set of Allocations for a Job by inserting Wait-Tacts. To ensure that the resulting increase in throughput-time is minimized, it will do so while minimizing the number of inserted Wait-Tacts. Given a Scenario $SO$, a Job $J$, a corresponding Start-Tact-ID ($STID$), a limit to the permissible Wait-Tacts $Lim_{WT}$, and a capacity limit relative to which to evaluate feasibility ($CapacityLimit$), the fragment is executed as follows.

To allow for an assignment of a Job with minimal throughput-time, a generic set of Splice-Points ($SPS$) with zero Wait-Tacts each is generated to accommodate $J$.

$$SPS = \rho_{SIdx,Count}(Sequence(max_{SIdx}(J \bowtie TSR)) \times 0) \qquad (5.106)$$

This now sets the stage for an iterative search for a feasible set of Allocations for the Job $J$. To minimize throughput-time the algorithm fragment will first generate a set of Allocations for $J$ using $SPS$ and the externally provided $STID$. In the first iteration the fully contracted form of the Tactsequence, which has no Wait-Tacts, is used. The Allocations in the generated set are then checked for feasibility.

$$ACS = genAllocations(J, SPS, STID) \qquad (5.107)$$
$$ACSFEAS = \{f \mid \forall a\, inACS(f.SIdx = a.SIdx) \wedge$$
$$(f.feas = feasibleScAcMerge(SO, a, CapacityLimit))\} \qquad (5.108)$$

This can yield three possible results:

1. **All Allocations feasible:** In this case the algorithm has found the set of Allocations with the shortest possible throughput-time and can terminate yielding the result.
2. **First Allocation infeasible:** Should this be the case, no insertion of Wait-Tacts will be able to render the Job's Allocations feasible. This is due to the fact that Wait-Tacts can only be inserted at Splice-Points; these are always located between two Segments. Inserting a Wait-Tact into the first Splice-Point thus has no influence on the feasibility of the first Segment. Since the Algorithm will be unable to find a feasible set of Allocations, it terminates yielding an empty set.

3. **Later Allocation infeasible:** Should an Allocation following the first be found to be infeasible, a Wait-Tact is inserted prior to the first infeasible Segment if this would not cause the maximum number of permissible Wait-Tacts $Lim_{WT}$ to be exceeded. After incrementing the number of Wait-Tacts for the appropriate Splice-Point in $SPS$, the algorithm regenerates all Allocations and resumes execution at Equation (5.107). Should an incrementation of the Wait-Tacts have been impossible to the aforementioned constraint, the algorithm fragment terminates yielding an empty set.

If the algorithm fragment completes without terminating as described in the three cases above, a feasible Allocation has been found and is contained in $ACS$. Should it terminate prematurely, $ACS$ is an empty set.

$$ACS = findFeasAllocations(SO, J, Lim_{WT}, STID, CapacityLimit) \qquad (5.109)$$

**Find Job with Largest Capacity Requirement on Least Utilized Constraint**

To ensure that the production system's constraints are properly exploited, an important task is the identification of Jobs, which make intensive use of the constrained machines. Thus this presented algorithm fragment serves to find the Job with the largest capacity requirement of the constraint with the lowest capacity utilization. Given a relation $(JL)$ containing Job-Tuples and a Scenario $(SO)$ the presented procedure consists of two steps: the identification of the constraint with the lowest utilization and the search for the Job with the largest Capacity Requirement on said constraint[291]

When determining constraint utilization, the observed window of time significantly influences the results. In this case it is expedient to focus on those Tacts which are directly affected by the scheduling operation. This limits the relevant Tacts to a window beginning with the Scheduling Horizon's $(SH)$ Start-Tact and ending with the last production Tact reachable by Jobs commencing processing on the latest possible Tact in $SH$. Given a Scenario $(SO)$, a Constraint Container $(CC)$ and set of Jobs to assign $(JS)$ the procedure is defined as follows.

First all relevant time-critical Jobs are isolated:

$$CritJobs = \sigma_{(PS_{SST} \leq LPST < (PS_{SST} + PS_{SH}))}(JL \bowtie LSTC). \qquad (5.110)$$

---

[291]The idea underlying this heuristic was introduces and successfully tested by Ziskoven in his work on single-Tactline-scheduling (cf. ZISKOVEN (2013) Methodik zur Gestaltung und Auftragseinplanung einer getakteten Fertigung im Werkzeugbau, pp. 135).

In this group, the latest potential production Tact (LPPT) can be deduced from the latest due-dates:

$$LPPT = max_{TID}(\sigma_{PIT < max_{DueDate}(CritJobs)}(Cal)). \tag{5.111}$$

The cumulative capacity supply (CCSPM) is calculated per Machine for the relevant Tacts:

$$CCSPM = \gamma_{MID,sum(TargetCapacity)}(\sigma_{PS_{SST} \leq TID \leq LPPT}(CS)). \tag{5.112}$$

Analogously the cumulative capacity requirements (CCSRM) are thus calculated as:

$$CRPM = \sigma_{PS_{SST} \leq TID+TSID-1 \leq LPPT}(SO \bowtie TSR \bowtie TSSR) \tag{5.113}$$

$$CCRPM = \gamma_{MID,sum(RequiredCapacity)}(CRPM). \tag{5.114}$$

Finally the constraint utilization (CU) can be calculated as:

$$CU_{interim} = \pi_{MID, \frac{sum(RequiredCapacity)}{sum(TargetCapacity)}}(CCSPM \underset{CCSPM.MID=CCRPM.MID}{\bowtie} CCRPM) \tag{5.115}$$

$$CU = \rho_{\frac{sum(RequiredCapacity)}{sum(TargetCapacity)}/LF}(CU_{interim}). \tag{5.116}$$

Now that the utilization for each constraint has been calculated, the constraint with the lowest utilization can easily be identified.

Initially all Capacity Requirements for Jobs in $JL$ are retrieved:

$$JLWCR = JL \bowtie TSR \bowtie TSSR \bowtie TSSC. \tag{5.117}$$

Now the load factors for the constraints (CU) can be combined with the Capacity Requirements in $JLWCR$ to allow for extraction of the Job with the largest capacity requirement on the constraint with lowest utilization ($J_{lcr}$)

$$CUwJL = CU \underset{CU.MID=CritJobsWCR.MID}{\bowtie} JLWCR \tag{5.118}$$

$$MinLF = min_{LF}(CUwJL) \tag{5.119}$$

$$MaxRC = max_{RequiredCapacity}(MinLF) \tag{5.120}$$

$$J_{lcr} = \pi_{JID}(first(MaxRC)) \tag{5.121}$$

Based on the description above a function $findJob_{LCR}(...)$ is defined, which receives a Scenario $SO$, a Constraint Container $CC$ and set of Jobs to assign $JL$ and returns the aforementioned Job.

$$J_{lcr} = findJob_{LCR}(SO, CC, JL) \tag{5.122}$$

**Find Job with most Even Capacity Consumption**

Aside from selecting the Job which will make the most use of the emptiest constraint, a second criterion is used primarily in the final phase of algorithm execution. The objective is to find the Job which will lead to the most even capacity consumption. Phrased differently; this is the Job which most effectively fills non-utilized Machine capacities along its respective Tactsequence and thus is the best fit in terms of utilizing available Processing Capacity. This serves the purpose of ensuring that the final "topping off" maximizeS the machine utilization.

The first step is the calculation of the currently available capacities. Since the Tacts in question and the calculation of the remaining capacity is similar to the calculation of machine utilization, the calculation of the relevant series of Tacts is identical and thus Equations (5.110) and (5.111) can be carried over as is. The description of this algorithm fragment picks up under the assumption that the aforementioned steps have been successfully completed. Given a relation $(JL)$ containing Job-Tuples and a Scenario $(SO)$ the procedure described below is comprised of two major steps: **the calculation of available Processing Capacities in a Scenario** and **the search for the Job, which represents the best fit**.

The selection of this "best fit" is dependent on the Tact in which the Job is to commence processing. Therefore this algorithm fragment requires the $TID$[292] which should represent the Tact in which the chosen Job commences production.

To calculate the availble Processing Capacities, initially the capacity supplies (CSPMT) are extracted per Machine and Tact

$$CSPMT = \sigma_{PS_{TID} \leq TID \leq LPPT}(CS). \tag{5.123}$$

Analogously the Capacity Requirements $(CSRMT_{interim})$ are extracted as

$$CRPMT_{assigned,interim} = \sigma_{PS_{SST} \leq TID+TSID-1 \leq LPPT}(SO \bowtie TSR \bowtie TSSR). \tag{5.124}$$

As was the case when checking the feasibility of a Scenario the $TID$s for Tactstation Requirements must be calculated based on the Start-Tact-ID $(TID)$ given at the Segment-level (refer to Equation 5.71). These $TID$s can be derived trivially by adding the Segment $TID$ in the Allocation and the $TSID$ in the Tactstation Requiremens and finally substracting 1.

---

[292]Since this fragment will be used in the context of the Scheduling Algorithm presented in later pages, the assumption that $SST \leq TID \leq LPPT$ definitely holds.

Thus $CRPMT_{corrected}$ is defined as

$$CRPMT_{assigned,interim2} = \pi_{MID,TID+TSID-1,RequiredCapacity}(CRPMT_{interim}) \quad (5.125)$$

$$CRPMT_{assigned,corrected} = \rho_{TID+TSID-1/TID}(CRPMT_{interim2})) \quad (5.126)$$

Now the required Processing Capacity used by allocated Jobs can be aggregated to $CCRPMT$.

$$CCRPMT = \gamma_{MID,TID,sum(RequiredCapacity)}(CRPMT_{corrected}) \quad (5.127)$$

The remaining available Processing Capacity for additional Allocations ($AVCAP$) can now be calculated:

$$AVCAP_{interim} = \pi_{MID,TID,TargetCapacity-RequiredCapacity}(CCRPMT \bowtie CSPMP) \quad (5.128)$$

$$AVCAP = \rho_{TargetCapacity-RequiredCapacity/AvailableCapacity}(AVCAP_{assigned,interim})). \quad (5.129)$$

The Job which would make the most uniform usage of the available capacity if it were to commence production in the Tact with the ID $STID$ must now be chosen from $JL$. During the final phase of the algorithm execution ("Topping Off"), the addition of Jobs to the Scenario is desirable but not essential. Thus in the interest of efficiency the algorithm only attempts to assign Jobs in their most "contracted" form - i.e. Jobs are only assigned without Wait-Tacts.

To allow for Job assignment with zero Wait-Tacts, a generic set of Splice-Points ($SPS$) is generated, which contains sufficient elements to accommodate the longest Tactsequence associated with a Job in $JL$.

$$SPS = \rho_{SIdx,Count}(Sequence(max_{SIdx}(JL \bowtie TSR)) \times 0) \quad (5.130)$$

In a next step the Allocations are generated for all Jobs inserting no Wait-Tacts.

$$ACS = \{a \mid \forall j \in JL(a = genAllocations(j, SPS, STID))\} \quad (5.131)$$

The effective Capacity Requirements are then derived for each Allocation in $ACS$ in a similar fashion as was the case above (refer to Equations (5.124–5.127))

$$CRPMT_{unassign,int} = ACS \bowtie SO \bowtie TSR \bowtie TSSR \quad (5.132)$$

$$CRPMT_{unassign,int2} = \pi_{JID,MID,TID+TSID-1,RequiredCapacity}(CRPMT_{unassign,int}) \quad (5.133)$$

$$CRPMT_{unassign,corrected} = \rho_{TID+TSID-1/TID}(CRPMT_{unassign,interim2})) \quad (5.134)$$

Now it is finally possible to calculate the relative consumed Processing Capacity ($RCCAP$) for each Job on each Tactsubstation along the Tactsequence:

$$RCCAP_{interim} = \pi_{JID,MID,TID,\frac{RequiredCapacity}{AvailableCapacity}} \left( AVCAP \bowtie CRPMT_{unassigned,corrected} \right)$$

$$\tag{5.135}$$

$$RCCAP = \rho_{RelativeCapacityConsumption/\frac{RequiredCapacity}{AvailableCapacity}} (CCAP_{interim})) \tag{5.136}$$

At this point the relation $RCCAP$ could potentially contain Jobs, where an assignment would not be feasible. Before proceeding, those Jobs which could not be manufactured with the remaining Production Capacities are eliminated from the relation.

$$RCCAP_{feasible} = \{t \mid (t \in RCCAP) \land (t.RelativeCapacityConsumption \leq 1)\}$$

$$\tag{5.137}$$

The next step is to determine the uniformity of relative capacity consumption for a Job across all Tactstations. Put simply a Job with a constant relative capacity consumption across all Tactsubstation would be a Job which perfectly fits the available Production Capacities. Thus the Job which represents the best possible fit maximizes the similarity of relative capacity consumption for each Tactstation. While various metrics exist, usage of the *standard deviation* yielded good results and thus will be applied in the context of this thesis. When applying the standard deviation to the problem at hand, the Job with the minimal standard deviation in its relative capacity consumption represents the best fit ($MECR$).

$$RCCAPSTDEV = \gamma_{JID,stddev(RelativeCapacityConsumption)}(RCCAP_{feasible}) \tag{5.138}$$

$$MECR = min_{stddev(RelativeCapacityConsumption)}(RCCAPSTDEV) \tag{5.139}$$

$$J_{mecr} = \pi_{JID}(MECR) \tag{5.140}$$

A function $findJob_{MECR}(...)$ is defined, which receives a Scenario $SO$, a Constraint Container $CC$, a set of Jobs to assign $JL$. It returns the Job, which best fits the available Processing Capacities.

$$J_{MECR} = findJob_{MECR}(SO, CC, JL) \tag{5.141}$$

## Phase 1 - Critical Job Allocation

Phase 1 of the presented Production Scheduling algorithm focuses on the scheduling of those Jobs which must commence production in the Scheduling Horizon in order to be completed by their respective due-dates. These will henceforth be referred to as **Critical Jobs** ($CJ$). The underlying idea is to first find a feasible Allocation for those Jobs which definitely need to be assigned to a Tact, to reduce the size and thus computational complexity of the optimization problem. This phase of the scheduling algorithm begins with a Scenario $SO$ containing no Allocations.

Due to the deterministic nature of synchronized manufacturing the total throughput-time for a Job can be determined prior to the commencement of production. This makes the calculation of the latest possible Start Tacts for each Job possible. During the execution of the preparatory steps the Latest Start-Tacts are determined and placed in the Latest Start-Tact Container $LSTC$ as described in Chapter 5.3.4.2. This allows for the simple identification of Critical Jobs by selecting those where $LPST$ lies in the Scheduling Horizon.

$$CJ = \sigma_{(PS_{SST} \leq LPST < (PS_{SST} + PS_{SH}))}(JL \bowtie LSTC) \qquad (5.142)$$

After having isolated the Critical Jobs, the objective is now to assign these to suitable Start-Tacts. Once an assignment has been made, the Job is removed from $CritJobs$. This is done by means of an iterative approach beginning with the first Tact in the Scheduling Horizon and ending with its final Tact. For the purpose of describing the steps performed during each iteration, the Tact being processed in the context of a given iteration is referred to as $CurrentTact$.

In a first step during each iteration, those Jobs in $CJ$ are chosen, which must commence production in $CurrentTact$.

$$CJ_{Tact} = \sigma_{LPST = CurrentTact.TID}(CJ) \qquad (5.143)$$

All Jobs in $CJ_{Tact}$ must be assigned to not violate the Jobs' due-dates. This entails that these Jobs cannot tolerate any Wait-Tacts since an inserted Wait-Tact is equivalent to a delayed Start-Tact by one Tact. Thus all Jobs must be assigned to the Start-Tact without Wait-Tacts.

To allow for a Job assignment without Wait-Tacts an appropriate set of Splice-Points ($SPS$) is generated to accommodate all Jobs in $CJ_{Tact}$.

$$SPS = \rho_{SIdx, Count}(Sequence(max_{SIdx}(CJ_{Tact} \bowtie TSR)) \times 0) \qquad (5.144)$$

Using the generated set of Splice-Points Allocations are now generated for all Jobs in $J_{Tact}$ and checked for feasibility. In this case $MaximumCapacity$ is used as the Capacity Limit, to ensure every effort is made to complete time-critical Jobs by using all available manufacturing resources.

$$ACS_{Tact} = \{a \mid j \in CJ_{Tact}$$

$$(a = genAllocations(j, SPS, CurrentTact.TID))\} \qquad (5.145)$$

$$ACSFEAS_{Tact} = \{f \mid a \in ACS_{Tact}$$

$$(f = feasibleScAcMerge(SO, a, MaximumCapacity))\} \quad (5.146)$$

Should $ACSFEAS_{Tact}$ contain one infeasible entry, not enough Production Capacities are available to allow for the timely completion of all Critical Jobs. Since the algorithm has no further options to resolve this situation, the human expert performing the production scheduling must intervene to resolve the situation and allow the algorithm to proceed. Specifically there are three possible actions to resolve the situation:

1. **Increase Processing Capacities:** A simple solution is to increase the maximum-capacity on the relevant bottlenecks which preclude the successful production of all Critical-Jobs. In practice this is can be achieved by using extra shifts or temporary inclusion of machinery otherwise not used by the synchronized production system.

2. **Postpone Jobs:** Another option is to extend the due-date of selected Jobs. This allows the algorithm to postpone the production of Jobs to later Tacts thus effectively also increasing the available production capacity for the given Job.

3. **Remove Jobs from Scheduling:** The final option is to reduce the number of Jobs to be scheduled until the available production capacities are sufficient to complete processing of all Critical Jobs. In practice this option is usually the least desirable, since this would result in removing Jobs from the synchronized production system and processing them externally.

Irrespective of the option chosen to resolve the issue of insufficient production capacity, the algorithm must restart at Equation (5.145) and attempt to assign the time-critical Jobs, which must commence production in *CurrentTact*, again. Once all entries in $ACSFEAS_{Tact}$ the Allocations generated for $CJ_{Tact}$ are added to the Scenario and removed from $CJ$.

$$SO = SO \cup ACS_{Tact} \qquad (5.147)$$
$$CJ = CJ - CJ_{Tact} \qquad (5.148)$$

The algorithm now assigns time-critical Jobs from $CJ$, which can start production on a Tact later than $CurrentTact$. These Jobs are grouped by $LPST$ and processed in ascending order. Within each group an attempt is made to assign the currently remaining Job with the largest largest capacity requirement on the least utilized constraint. Restricting search to the group ensures that Jobs are processed in descending order of their "criticality". If an assignment is feasible, the resulting Allocations are added to the Scenario and the Job is removed from $CJ$. Regardless of the success, successive attempts are made to assign the Job which are deemed to next-best according to the aforementioned criterion.

To this end an entity containing all remaining Latest Possible Start-Tacts is extracted and deduplicated.

$$LPST_{remaining} = \delta(\pi_{LPST}(CJ)) \qquad (5.149)$$

Select the element from $LPST_{remaining}$ where the $LPST$ is minimal and remove $LPST_{current}$ from $LPST_{remaining}$.

$$LPST_{current} = min(LPST_{remaining}) \qquad (5.150)$$
$$LPST_{remaining} = LPST_{remaining} - LPST_{current} \qquad (5.151)$$

Next choose all Jobs with $LPST = LPST_{current}$.

$$CJ_{curLPST} = \sigma_{LPST=LPST_{current}}(CJ) \qquad (5.152)$$

Now find the Job with the largest capacity requirement on the least utilized constraint ($J_{LCR}$) in $CJ_{curLPST}$.

$$J_{LCR} = findJob_{LCR}(SO, Target - Capacity, CJ_{curLPST}) \qquad (5.153)$$

Having now identified the preferred assignment candidate ($J_{LCR}$) the attempt is made to find a suitable allocation. Since it is not critical that $J_{LCR}$ to $CurrentTact$, $TargetCapacity$ is used as capacity limit. It is important to note, that during this assignment, the number of permissible Wait-Tacts is not necessarily zero. Now Jobs with a Latest Possible Start-Tact which is later than $Current-Tact$ are being chosen from. The maximum number of permissible Start-Tacts is now bounded by both the $PS_{MNWT}$ parameter defined in Chapter 5.3.4.1 and the "slack" between $CurrentTact$ and $LPST_{current}$.

$$ACS_{J_{LCR}} = findFeasAllocations(SO, J_{LCR},$$
$$min(\{PS_{MNWT}, CurrentTact.TID - LPST_{current}\}),$$
$$CurrentTact.TID, TargetCapacity)If$$

$$(5.154)$$

$ACS_{J_{LCR}}$ is not an empty set, a feasible assignment has been found and the Allocations yielded by $findFeasAllocations$ can safely be added to $SO$. Since the Jobs could theoretically be moved to other Tacts in ensuing Algorithm-Phases their Status Flag is set to $reversible$ (refer to Chapter 5.3.2).

$$SO = SO \cup ACS_{J_{LCR}} \qquad\qquad (5.155)$$

$$CJ_{curLPST} = CJ_{curLPST} - J_{LCR} \qquad\qquad (5.156)$$

If $CJ_{curLPST}$ is a non-empty set, the algorithm continues with Equation (5.150).

Once all groups have been processed, the assignment of Jobs to $CurrentTact$ is complete. If $CurrentTact$ is not the last Tact in the Scheduling Horizon ($SH$), $CurrentTact$ is incremented and the procedure is once again repeated beginning with Equation (5.145). If the end of $SH$ is reached, all Jobs which must commence production in $SH$ have been successfully allocated and thus Phase-1 of the algorithm is complete.

**Phase 2 - Rigging Optimization**

The Rigging Optimization is executed after all Critical Jobs have been successfully allocated. In this second phase of the scheduling algorithm the objective is to coalesce components, which have identical $RiggingKeys$ on a Machine, to receive processing on this Machine during the same Tact. Since a $RiggingKey$ encodes the setup prerequisites necessary to commence production on a Machine, this serves to reduce the number of rigging operations necessary in a Tact. In turn, this increases the proportion of productive time during a Tact. While the heuristic does not perform an exhaustive search of the solution space, it nevertheless

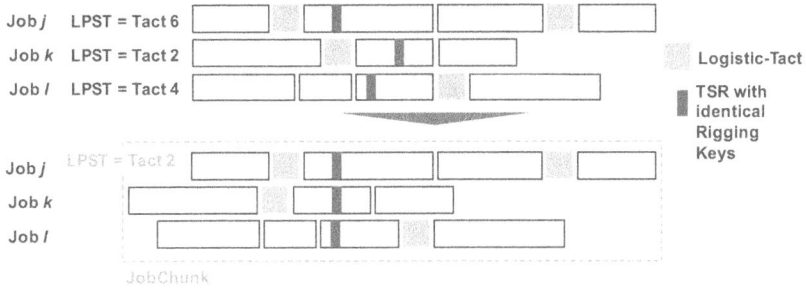

Figure 5.6: Generation of a Job-Chunk for Jobs j, k and l

yields good results in industrial practice as is shown in Chapter 6. The proposed procedure is described in detail henceforth.

Initially all Allocations are removed from the Scenario and all associated Critical Jobs are extracted for reallocation:

$$CJ_{critical} = \pi_{JID}(SO) \bowtie JP. \tag{5.157}$$

$$\tag{5.158}$$

Afterwards all Jobs associated with the removed Allocations are grouped in accordance with contained rigging keys. This grouping operation yields so-called **Job-Chunks** ($JC$). A Job-Chunk represents a set of Allocations for two or more Jobs with identical Rigging-Keys. If this set of Allocations is added to the Scenario, it will ensure that the referenced Jobs arrive at a certain Machine for processing in the same Tact. Figure 5.6 shows an example, where the Jobs j, k and l are grouped and aligned to form a Job-Chunk with a resulting $LPST$ of 2. Thus each Job-Chunk contains a set of Allocations which ensures that the "chunked" Jobs are processed on the affected machine in the same Tact. Job-Chunks are placed in a **Job-Chunk Container** ($JCC$).

To generate the set of Job-Chunks a series of steps are necessary. Job-Chunks are groupings of **Jobs** with identical **Rigging-Keys** on specific Machines, which are scheduled for processing in a specific **Tact**. Thus one approach to generating all possible Job-Chunks is to first generate all feasible possibilities of commencing production for all Critical Jobs[293].

$$TxJ = \sigma_{TID \leq LPST}(SH \times CJ_{critical} \bowtie LSTC) \tag{5.159}$$

Now the corresponding Allocations with zero Wait-Tacts are generated for all possible Start-Tacts in TxJ

$$SPS = \rho_{SIdx,Count}(Sequence(max_{SIdx}(CJ_{critical} \bowtie TSR)) \times 0) \tag{5.160}$$

$$ACS = \{a \mid (t \in TxJ, j \in CJ_{critical})((t.JID = j.JID) \wedge$$

$$(a = genAllocations(j, SPS, t.TID)))\}. \tag{5.161}$$

The individual Allocations' Start-Tacts can be used to calculate the Tacts in which a Job reaches a certain Machine ($JMT$).

$$JMT_{interim} = ACS \bowtie SO \bowtie TSR \bowtie TSSR \tag{5.162}$$

$$JMT_{interim2} = \pi_{JID,MID,TID+TSID-1,RequiredCapacity}JMT_{interim}) \tag{5.163}$$

$$JMT = \rho_{TID+TSID-1/IntersecTID}(JMT_{interim2})) \tag{5.164}$$

To ensure only relevant combinations of Job, Machine and Tact are considered, entries without a Rigging-Key are purged from $JMT$.

$$JMT_{cleaned} = \sigma_{RiggingKey \neq EMPTY}(JMT) \tag{5.165}$$

By aggregating on the combination of RiggingKey, Machine and Tact it is possible to count the number of Jobs with a given Rigging-Key, which arrive on a Machine during a specific Tact. Since the algorithm is searching for groups of Jobs, which could arrive simultaneously, only those candidates are of interest, where two or more Jobs are involved.

$$JC_{interim} = \gamma_{MID,IntersecTID,RiggingKey,count(*)}(JMT_{cleaned}) \tag{5.166}$$

$$JC_{candidates} = \sigma_{count(*)>1}(JC_{candidates,full}) \tag{5.167}$$

In a final step the Job-Chunk candidates are converted into Job-Chunks ($JC$)

$$JCC = \{t \mid (c \in JC_{candidates})(t.Allocations =$$

$$\sigma_{(MID=t.MID) \wedge (IntersecTID=t.IntersecTID) \wedge (RiggingKey=t.RiggingKey)}(JMT_{cleaned}) \wedge$$

$$t.JobCount = c.count(*))\} \tag{5.168}$$

The ensuing Rigging Optimization is comprised of two sequential stages, which are executed repeatedly until completion. In the first stage all remaining *reversible* Allocations are removed from the Scenario.[294] Next the largest Job-Chunk, i.e. the Job-Chunk containing the greatest number of items is identified ($MaxJC$). The contained Allocations are added to the Scenario and the Job-Chunk is removed from $JCC$.[295]

---

[293]In accordance with requirements expressed by industrial experts, Rigging Optimization generally does not justify an extension of a Job's throughput time. Therefore Jobs are pnly scheduled without Wait-Tacts during Job-Chunk generation. This also reduces the problem size and the associated complexity

[294]During the first iteration the scenario is already empty at this point.

[295]At this point the added Allocations have the status flag reversible.

$$SO = SO - \sigma_{Status=reversible}(SO) \tag{5.169}$$

$$MaxJC = max_{JobCount}(JCC) \tag{5.170}$$

$$SO = SO \cup MaxJC.Allocations \tag{5.171}$$

$$JCC = JCC - MaxJCCJ_{Rest} \qquad = CJ_{critical} - (\delta(\pi_{JID}(MaxJC.Allocations)) \bowtie JP) \tag{5.172}$$

In the second stage an attempt is made to reallocate all remaining unallocated Jobs in a fashion similar to the first phase of the presented algorithm chapter 5.3.5.2. In the interest of succinctness only the modifications to this procedure are described at this point.

- In Equation (5.143) the set of Jobs to be assigned $CJ$ is replaced with $CJ_{Rest}$
- The Capacity limit *MaximumCapacity* is lowered to TargetCapacity in Equation (5.146) since the Rigging optimization should only be executed, if it is possible without causing the Target-Capacity to be exceeded
- No manual intervention of the user is permissible, since this would imply that no feasible allocation was possible due to the assigning of the Job-Chunk

This process can complete with one of two results:

1. **Allocation Feasible:** Should the reallocation prove feasible this means that the allocation of the Job-Chunk can be performed without violating the due-dates of Jobs not in the Chunk. To preserve this viable grouping of Jobs the status of the associated Allocations is set to *fixed*. By fixing these Allocations, the referenced Jobs are now scheduled for production and will not be moved during subsequent algorithm execution. Thus these Jobs are removed from all Job-Chunks in the $JCC$. Should Job-Chunks be reduced to containing only a single Job by this operation, they too are purged from the $JCC$ in this step. If there are Job-Chunks remaining in $JCC$ after completion of these steps, the algorithm resumes execution at the beginning of the first stage.

2. **Allocation Not Feasible:** In this case adding the grouping of Jobs to the Scenario precludes a successful allocation of all Critical Jobs. Since all Jobs were successfully allocated in phase 1 of the scheduling algorithm, this must be due to Allocations contained in the Job-Chunk. Therefore the problematic Job-Chunk is divided into two Sub-Chunks of ideally equal size. The resulting Sub-Chunks which reference two or more Jobs are then added to $JCC$. If the $JCC$ is non-empty, the algorithm resumes execution in stage one or terminates otherwise.

After running to completion, the algorithm has iteratively attempted to rearrange the Critical-Jobs allocated in phase 1 to reduce the necessary Machine setup time. This marks the point at which the Status Flag of all Allocations in the Scenario is set to *fixed*. Thus the scheduling of Critical-Jobs will not be modified in the final phase of the algorithm.

$$\forall a \in SO$$

$$a.Status = fixed \qquad (5.173)$$

**Phase 3 - Topping Off**

The third and final phase of the algorithm maximizes Machine utilization by scheduling Jobs which are not yet considered time-critical. To ensure that Jobs are not produced too far in advance, the algorithm chooses among Jobs which are considered critical in an ***Expanded Scheduling Horizon*** $(ESH)$. ESH includes all Tacts in the Tact Horizon beginning with the first Tact following the Scheduling End Tact $SET$ and ending with $SET + 1 + (ESHPS_{SH} * (PS_{SHEF} - 1))$. It is important to note, however, that while the Expanded Scheduling Horizon provides Jobs with which remaining production capacities are filled, all assignments are performed in the "non-expanded" Scheduling Horizon. Since the procedure is similar to the Allocation of critical Jobs, the subsequent description is kept short and will only be detailed in those areas where the two approaches differ.

Using the Latest Start-Tacts for each Job calculated in the preparatory steps the Jobs are determined which must commence production in the ESH.

$$ESHCJ = \sigma_{(SET+1) \leq LPST \leq (ESHPS_{SH} * (PS_{SHEF} - 1))}(JL \bowtie LSTC) \qquad (5.174)$$

After isolating Jobs critical to the ESH, these are assigned to Tacts in the SH. This is once achieved by iterating over Tacts beginning with the first Tact in the Scheduling Horizon and ending with its final Tact. $CurrentTact$ in this case references the Tact, which is the focus of the current iteration. During each iteration the Job with the most even capacity consumption if it were to commence production in $CurrentTact$ is chosen and assigned to the Scenario if feasible. This process is repeated until all Jobs from ESHCJ have been processed. A grouping of Jobs according to their criticality is not considered necessary, since none of the Jobs processed in this Phase need to start production in SH. The object is rather to choose those

Jobs for the assignment which represent the best possible fit for the remaining Processing Capacities.

At the beginning of every iteration, all Jobs remaining in $ESHCJ$ are transferred to the relation $ESHCJ_{Tact}$:

$$ESHCJ_{Tact} = ESHCJ. \tag{5.175}$$

The next step is to find the Job with the most even capacity consumption if it were to commence production in $CurrentTact$.

$$J_{MECR} = findJob_{MECR}(SO, Target - Capacity, ESHCJ_{Tact}) \tag{5.176}$$

Having now identified the preferred assignment candidate ($J_{MECR}$) an attempt is made to assign the Job in its most contracted form since Topping-Off should not lead to the unnecessary insertion of Wait-Tacts. Since assigning $J_{MECR}$ is an optional measure designed to fill potential gaps between Machine utilization and $TargetCapacity$, this limit is used as the effective Capacity Limit.

$$ACS_{J_{MECR}} = findFeasAllocations(SO, J_{LCR}, 0, CurrentTact.TID, TargetCapacity) \tag{5.177}$$

If $ACS_{J_{MECR}}$ is not an empty set, yielded by $findFeasAllocations$ can safely be added to $SO$ (refer to Chapter 5.3.2).

$$SO = SO \cup ACS_{J_{LCR}} \tag{5.178}$$

$$ESHCJ = ESHCJ - J_{MECR} \tag{5.179}$$

In any case remove $J_{MECR}$ from $ESHCJ_{Tact}$

$$ESHCJ_{Tact} = ESHCJ_{Tact} - J_{MECR}. \tag{5.180}$$

If $ESHCJ_{Tact}$ is a non-empty set, the algorithm continues with Equation (5.176).

Once all Jobs in $ESHCJ_{Tact}$ have been processed, the assignment of Jobs to $CurrentTact$ is complete. If $CurrentTact$ is not the last Tact in the Scheduling Horizon ($SH$), $CurrentTact$ is incremented and the procedure is once again repeated beginning with Equation (5.175). Once the end of $SH$ is reached, the Jobs from the Expanded Scheduling-Horizon which represent the best fit for the current Scheduling-Horizon have been added to the Scenario, where possible. In a final step all Allocations are marked as fixed:

$$\forall a \in SO$$

$$a.Status = fixed \tag{5.181}$$

# Chapter 6

# Evaluation

In accordance with ULRICH's demands, this chapter returns to the practical application from which the research process originated. The objective in this Chapter is to evaluate whether the developed algorithms satisfy the requirements of the practical application formulated. The underlying hypothesis is that practical applicability of the presented algorithms is given. This hypothesis will be considered valid until evidence to the contrary is discovered. The process of evaluation, designed to systematically examine practical applicability is divided into two major stages:

1. **Theoretical Expert Validation (Chapter 6.1)**

   A panel of industrial experts in the area of synchronized production was consulted iteratively throughout the various stages of the research process. This allowed for a continuous theoretical validation of algorithm input, structure, flow and output.

2. **Practical Case-Study Validation (Chapter 6.2)**

   After completing the design of both the Tactsequence Mapping and Production Scheduling algorithms, a practical validation was conducted by implementing the algorithm and applying it in several industrial case-studies.

The following Chapters will provide a detailed overview of the activities in both stages of validation.

## Theoretical Evaluation

To help guide the research process outlined in Figure 1.5, the work underlying this thesis was subjected to continuous theoretical evaluation by a panel of industrial experts on the topic of synchronized production. More specifically the panel consisted of the directors and the upper management of internal Tool-Manufacturing divisions in three German automotive first tier suppliers. For each company, the management consisted of the heads of product development, production engineering and production. All three companies had previously implemented a synchronized production system as described in Chapter 2.3.3 and had been practicing manual production planning since its introduction. Thus the involved parties had a profound understanding of the production system. Furthermore members of the panel were well acquainted with the requirements arising from day-to-day operations and had developed best-practices for their manual operations.

Theoretical evaluation activities focused on three stages of ULRICH's process of applied research, each of which will now be summarized.

**Ascertainment of Practical Problems**
The author first became involved with synchronized make-to-order production at the conclusion of InSynchroPro II research project. These experiences in combination with desk research of the problem allowed for the first identification of practical problems in the area of production planning. To verify whether the problems had been identified correctly, two workshops with the industrial expert panel and three research associates were organized. In these workshops two key areas where further research was necessary were established: **Tactsequence Mapping and Production Scheduling**.

Regarding the former, experts affirmed that the absence of a methodology to bridge the gap between the output generated by production engineering and the input necessary for Production Scheduling causes issues in day-to-day operations. One company had integrated the generation of Tactsequences into the process of production engineering and thus generated the Process Step Series for individual Jobs based on the available Tactlines on the Factory Floor. Since this process was executed manually, the quality of the generated Tactsequences varied over time. Dips in Tactsequence-Quality were usually due to adjustments in the Tactline-Layout and the introduction of less experienced employees into the process. Furthermore, the process was deemed considerably more time-consuming than conventional production engineering, where the primary objective is to determine a short and efficient series of technological steps to produce a product.

Regarding Production Scheduling the experts confirmed that ZISKOVEN's scheduling algorithm produces very good results for Tactline-Layouts consisting of only a single Tactline. The sequential iterative application of the algorithm to multiple Tactlines, however, proved impractical in real world applications. All of the selected companies had moved from a single-line setup to a multiple-line setup to benefit from the inherent coverage and efficiency. As an interim solution all companies had developed tools which allowed for manual scheduling. All parties agreed that the prevalence of scheduling errors was due to the high inherent complexity. This furthermore rendered scheduling excessively time-intensive and manual Rigging Optimization completely infeasible. Thus the hypothesized industrial demand for a scheduling algorithm capable of handling multiple concatenated and intersecting Tactlines was validated.

**Examination of Relevant Practical Application**

To ensure that research was proceeding under the correct premises, six individual expert interviews followed by a large workshop including all experts were conducted. The objective in the preliminary interviews was to ascertain that the concept under development would be applicable to all three companies. Although all three of them had implemented a synchronized production system, each company had focused on different aspects when performing the roll-out. One company had invested into Rigging Optimization, another had set up a flexible system consisting of nine intersecting, concatenated Tactlines while the third had performed extensive integration of external suppliers into its processing system.

Thus the interviews were used to systematically map the current production planning processes at each of the three companies and perform SIPOC analyses for the Tactsequence Mapping and Production Scheduling steps. Since the practical validation was planned to include the implementation of the developed algorithms into software, potential data-formats and available IT-infrastructure were examined. The focus during this phase was on the middle management and shop-floor levels.

After consolidating all information gathered in the interviews, a management workshop was convened, where a consolidated conceptual outline was presented and deemed applicable. Furthermore management priorities regarding *adherence to due dates*, *throughput-time* and *Rigging Optimization* were determined. While the relative importance of these issues was established unanimously for Production Scheduling, preferences diverged regarding Tactsequence Mapping. For Tactsequence Mapping some managers valued the logistical simplification inherent in maximizing the Allocation of jobs to Tactlines. To achieve a higher proportion of Jobs mapped to Tactlines, these considered a moderate extension of throughput-time acceptable. Due to external pressure the priorities at one of the companies lay primarily on a minimization of throughput-time. This divergence is reflected in the set of parameters allowing for extensive customization of Tactsequence Mapping.

**Derivation of Design-Rules**

The final stage of theoretical evaluation was performed in a single expert workshop, shortly before the design of the algorithms was completed. The objective at this point was the logical validation of the algorithm's micro-structure and discussion of industrial best practices. Since the Production Scheduling algorithm is a heuristic and therefore the results depend heavily on the manner in which steps are executed, this was the primary focus of the logical validation. The overwhelming consensus was that the algorithm not only incorporates the industrial best-practices, but can also be readily understood and thus trusted by those responsible for Production Scheduling in daily practice. The experts used the opportunity to stress the vital importance of comprehensibility in light of the necessity to perform manual interventions when the algorithm cannot find a solution on its own. Aside from this aspect, the workshop was used to derive an implementation plan for the practical validation. In the interest of conciseness, a description is included in the next Chapter.

# Practical Evaluation

## Implementation Planning

The implementation planning defined the steps necessary to successfully implement Tactsequence Mapping and Production Scheduling processes based on the algorithm presented in Chapter 5. The implementation process was comprised of three steps which are described below.

1. **Status Quo Process Mapping**

   In a first step the current process of mapping Jobs to Tactlines and scheduling of Jobs was recorded. To this end, all major Process Steps were identified. The responsible and executing persons as well as the involved stakeholders were then identified and interviewed. Furthermore the content and format of all information flowing to and from these Process Steps was ascertained. An example of information flowing into the Tactsequence Mapping or Production Scheduling was Job-related data provided by a company's *Enterprise Resource Management System* $(ERP)$[296]. Examples of information flowing outward were the Job-Schedules or printouts which specify the Jobs which were to be loaded onto which Tactpallet by logistics.

---

[296]An ERP-System is a software system developed to support and integrate the various business processes such as sales, product design and development, manufacturing, inventory management, human resources, finance and accounting.

2. **Future Process Modelling**

The second step was the modelling of the future processes based on the information collected in the previous step. To allow for a smooth implementation, each Process Step and flow information was analyzed regarding its necessity. Those elements from the old process deemed necessary were integrated into a rough draft of the new process, which took the strategic objectives and capabilities of the individual companies into account. The detailed definition of the new mapping and scheduling processes was performed in collaboration with the responsible persons and stakeholders of the old process.

Since the initial situation was very different in the group of companies participating in this evaluation, the future processes were quite different as well. Thus, while the algorithm described in Chapter 5 was applicable to all involved companies, the practical application varied substantially. Examples of influencing factors are: the availability of IT-systems, the organization of logistics processes or the transfer of information generated by the planning process to the shop-floor level.

3. **Implementation Workshops**

The third and final step was a series of implementation workshops in order to prototypically test the process designed in the previous step. The workshops were executed immediately prior to the final implementation of the new process and represented the final rehearsal. As a preparation for this workshop all information necessary for the execution of both the old and the new process was prepared for the workshop day. The implementation workshop consisted of two steps. Initially the old process (either Tactsequence Mapping or Production Scheduling) was executed in its old, manual form. Following this, the new process was executed using identical information reformatted to be compatible with the new process.

The steps outlined above were executed and allowed companies to successfully prepare for the practical evaluation of the algorithm in the case-studies presented in following two Chapters.

# Case-Study: Tactsequence Mapping

To evaluate the behavior and performance of the Tactsequence Mapping algorithm, the tool manufacturing facility to which synchronized manufacturing had been introduced three years earlier was chosen. Since the installation of the production system, the Tactline-Layout had grown from a single pilot-Tactline to a total of 6 Tactlines, which incorporated a total of 51 Machines on the factory-floor. Each Tactline contained between 3 and 19 Tactstations or between 3 and 32 Tactsubstations. A set of 243 Jobs, which amounts to approximately a

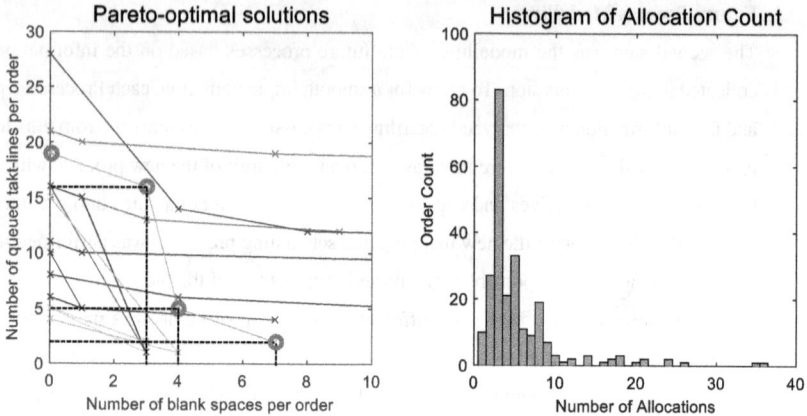

Figure 6.1: Pareto sets for selected Jobs (left) and Distribution of the number of Pareto optimal Tactsequences per Job (right)

week's worth of Jobs for this supplier was chosen as the evaluation set of Jobs. Preparatory interviews with the other two tool manufacturers revealed similar average Job throughput. All of the selected Jobs had previously been mapped manually. This allowed for the appraisal of the quality of results generated by the algorithm. The Jobs' routings contained a maximum, average and median of 35, 17.3 and 16 Process Steps per order respectively.

The algorithm was implemented as a single-threaded C# application and was executed on an Intel i7 processor with 8GB of memory. With this setup, the complete Tactsequence Mapping executing required 1346 milliseconds to completely process all Jobs. Thus using contemporary hardware, the prototypical algorithm runs with acceptable performance for day-to-day use. Since the implemented Branch-and-Bound algorithm is inherently parallelizable, performance could be increased further with relative ease.

The left part of Figure 6.1 shows a two-dimensional simplification of the three-dimensional Pareto set generated during algorithm execution. The ordinate displays the sum the number of Sequence Segments per job[297] while the abscissa contains the number of Idle-Tacts inserted. The colored lines shown in the graph each represent the Pareto set for a single Job. The green curve with the red highlights represents a Pareto front containing four Pareto optimal Tactsequences for a single Job. Results range from an Allocation without blank-spaces and 19 queued lines up to an Allocation with seven blank spaces and only two required processing

---

[297]The number of Sequence Segments is equal to the sum the number of Tactlines and number of isolated processing steps.

lines. With increasing numbers of Idle-Tacts the algorithm is able to perform a less exact matching of Process Step subsequences to Tactlines and can thus represent the entire Process Step Series with Tactlines rather than isolated processing steps. Thus this figure illustrates the positive effect of increasing the number of Idle-Tacts on the number of logistics operations. The right side of Figure 6.1 shows the distribution of the total number of Pareto optimal Tactsequences for all Jobs in the Test-set. Closer examination reveals that a half of all Jobs have three or fewer candidate Tactsequences at algorithm completion, whereas there are 4 outliers, which have up to 35 elements in their Pareto sets. This is an indication of the overall effectiveness of the Pareto based pruning during each algorithm iteration.

As shown above, increasing the permissible number of Idle-Tacts leads to a reduction in the number of necessary logistics operations. This positive side-effect, however, comes at the cost of increased throughput-time. Figure 6.2 shows the effects of increasing $TSM_{ATSD}$ on both the average throughput-time and the number of Segments in the Tactsequence. The most extreme case of allowing for not limiting the number Idle-Tacts reduces the number of Tactlines and isolated manufacturing steps in the resulting Tactsequence by approx. 53% and 70% respectively. The sharper drop in isolated manufacturing steps can be attributed to the fact that it is easier to merge a single step into a Tactline rather than the several steps covered by a Tactline. While this extreme measure results in throughput-time penalty of approx. 45%, a limit of 10 Idle-Tacts only results in a penalty of approx. 13% and at the same time effectively still almost cuts the number of logistic operations in half. Thus accepting 10 Idle-Tacts represents a sweet-spot in Figure 6.2. Similar evaluations in each of the other tool manufacturers manufacturing facilities similarly yielded sweet spots. This highlights the necessity of properly determining the parameters for Tactsequence Mapping in accordance with the Tactline setup and the processed Job-Spectrum (refer to Chapter 6.2.1).

Manual Allocation of the Jobs had averaged 5.18 Tactlines, 3.73 isolated manufacturing steps and a throughput-time of 21.44 Tacts. Thus by virtue of the production engineer's expertise manual Allocation yields results very close to the optimal values generated with the Tactsequence Mapping algorithm. One caveat however remains: since the necessary data could not be obtained to evaluate the Tactsequences on a monetary basis in Mapping Completion (refer to Chapter 5.2.2.3), the results of a monetary optimization could differ from the results obtained during this evaluation. Thus the results generated when taking inventory and logistics costs into account could differ significantly from the manual Tactsequence Mappings' results.

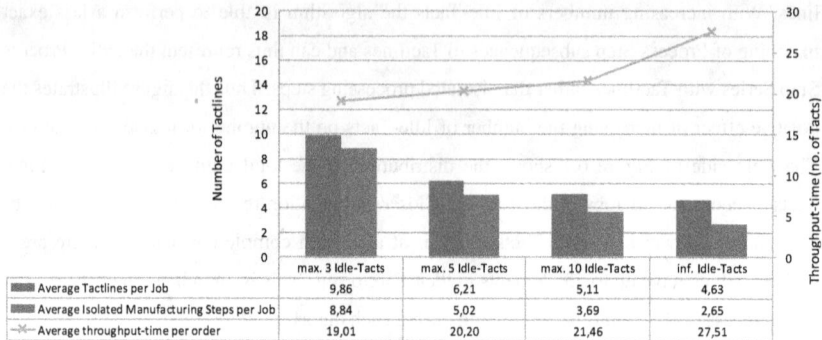

| | max. 3 Idle-Tacts | max. 5 Idle-Tacts | max. 10 Idle-Tacts | inf. Idle-Tacts |
|---|---|---|---|---|
| ▓ Average Tactlines per Job | 9,86 | 6,21 | 5,11 | 4,63 |
| ▓ Average Isolated Manufacturing Steps per Job | 8,84 | 5,02 | 3,69 | 2,65 |
| ─✕─ Average throughput-time per order | 19,01 | 20,20 | 21,46 | 27,51 |

Figure 6.2: Average number of Tactlines, isolated manufacturing steps and throughput time depending on the maximum number of allowed Idle-Tacts

## Case-Study: Production Scheduling

After successful execution of the Implementation Plan described above, a software-prototype of the Production Scheduling algorithm was evaluated on the factory floor. To validate two fundamental objectives guiding the algorithm's design, this Chapter focuses on the evaluation of constraint utilization and Rigging Optimization. Since the scheduling of Jobs to be completed in-time was incorporated into the algorithm as a hard constraint, this requirement was always fulfilled in all generated scheduling results and will not be elaborated further in this Chapter.

To evaluate the algorithm's ability to successfully exploit a production system's constraints a complex production setup consisting of 31 Machines and 7 concatenated intersecting Tactlines was chosen. A period of 25 production Tacts was chosen as the evaluation timespan. Since the company had opted to use a Tact-Frequency of $1\frac{Tact}{weekday}$, the regarded evaluation timespan is equivalent to five full production weeks. During this period a total of 1623 unique Jobs were processed on the selected Machines. To ensure the representativity of the results, the software was first introduced and integrated as the sole means of scheduling production a full 2 months prior to commencing evaluation. Thus the scheduling algorithm can be considered to be working in a steady state during the entire evaluated timespan.

Figure 6.3 shows utilization of constraints relative to the predetermined maximum-capacity during the evaluation timespan. In this production setup Constraint Identification (refer to Chapter 5.3.4.3) classified the Turning-, Nitriding- and Milling Machines to be elements constraining the overall output of the production system. Over the regarded period, the

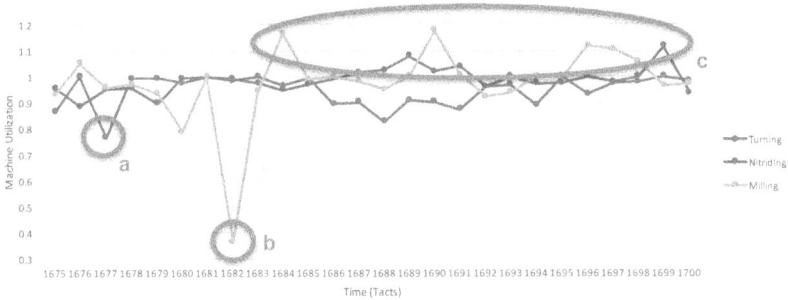

Figure 6.3: Machine utilization of the three Constraint-Machines during the evaluation timespan

algorithm was able to select and schedule Jobs and thus achieve the following Machine utilizations:

|  | Turning-Machine | Nitriding-Machine | Milling-Machine |
|---|---|---|---|
| Average Utilization | 0.94 | 0.98 | 0.97 |
| Std-Dev of Utilization | 0.06 | 0.04 | 0.15 |

In general, an average Machine utilization of more than 94% with a standard deviation of approximately 5% was achieved for all constraints. The industrial experts considered this to be better than the level of Machine utilization achieved prior to the introduction of synchronized production. This shows that the continuous deterministic flow of materials can allow for efficient use of the available constraint resources. The dip in Machine utilization marked as 'a' in Figure 6.3, however, also shows a disadvantage of a synchronized flow. The reduced utilization at marking 'a' was caused by a problem in a drilling process, which preceded the turning machine. Thus the problem in early Process Steps can send ripples of reduced efficiency through the entire production system. The massive drop in utilization marked by 'b' is due to a breakdown of the milling machine. This is also reflected in the high standard deviation in the utilization (15%). This reduced the available Machine time in the Tact significantly and led to the removal of more than half of the Jobs from their Tactpallets. After negotiations with the affected customers, Jobs were either processed by external suppliers or handled by increasing Machine utilization to more than 100 percent near marking 'c'.[298]

---

[298]This "overbooking" of Machines is possible due to two effects manifesting themselves in practice. Firstly all processing requirements included in Production Scheduling are estimates, which have a margin for error. In practice workers at Machine often regard the occasional handling of an overbooked Tactpallet as a personal challenge if it is indeed possible to complete the Jobs in a timely manner. A second possibility to handle excessive capacity requirements in a Tact is to schedule extra overtime between production Tacts.

The second focus when evaluating Production Scheduling was the examination of the Rigging Optimization. Since the previous production setup did not use Rigging Keys, a second setup was chosen. Since favor this case a small setup containing of a single Tactline consisting of five Tactstation was chosen. This setup is shown in the middle of Figure 6.4. Here the set of Turning-Machines (Tactstation 2) consisted of two Tactsubstations: one featuring a relatively small and the other a relatively large turning lengths. Due to Job characteristics such as the part-size and required precision, Jobs could either be processed exclusively on the small Machine (28%), large Machine (47%) or were compatible with both (47%). The regarded Scheduling Horizon encompassed 10 Tacts. Scheduling passes were first performed manually and then a scheduling run was performed using the developed software prototype.

The tables and charts at the top and bottom of the Figure show the distribution of the Jobs scheduled for Tactstation 2.1 and 2.2 and the resulting absolute Machine utilization. One important detail is that all Rigging Keys were successfully consolidated on either the large or small turning Machine. Furthermore no Rigging Key was divided into more than two groups for processing in different Tacts. In addition 22 of 32 Rigging Keys were successfully consolidated for processing in a single Tact, while the remaining 10 Rigging Keys were subdivided into a maximum of two groups each. With a total required processing capacity of 100.94 hours, the time savings amounted to a total approximately 4.51 hours or a 4.47% of additional productive processing time on the optimized Machines. The average relative Machine utilizations were 96.5% and 97.1% on the small and large Machines respectively. Comparisons with the manual planning results showed moderate increases in Machine utilization of 2.8% and 3.1%. In light of the fact that previously Machine utilization was a high priority objective during manual scheduling, this result was considered surprising by experts on the factory floor. Rigging Optimization was however previously considered to be too complex to perform manually. Thus the minor Rigging Optimizations amounting to 0.21 hours of productive time, which were present in manual scheduling results were the result of chance. In this area of Rigging Optimization, the algorithm produced significantly better results than manual scheduling.

# Summary

By combining a theoretical and a practical evaluation and performing both continuously during the research underlying this thesis, an algorithm was developed which produces high-quality results in industrial settings. This evaluation showed that the scheduling

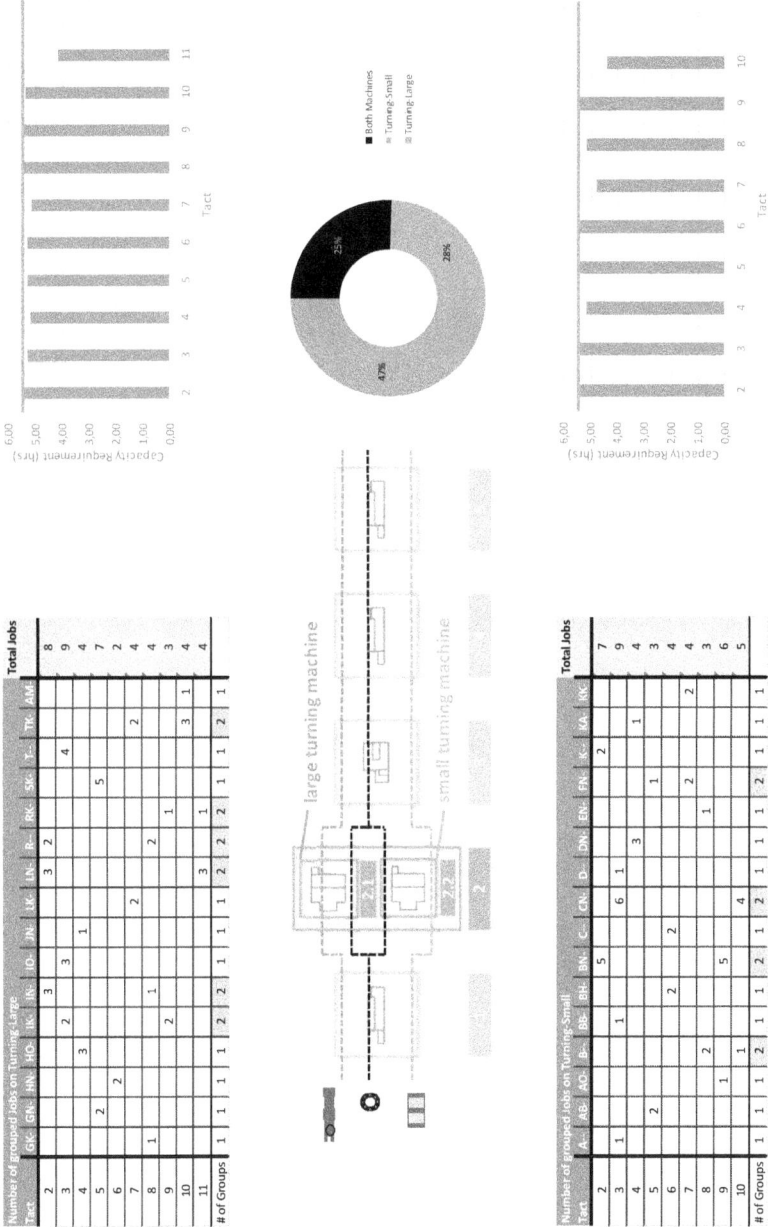

Figure 6.4: Property profile of the diverse library compared to the compound pool.

algorithm presented in Chapter 5 produced as good or better mapping and scheduling results than were possible in the predominantly manual status quo.

During the practical evaluation the expert panel deemed the algorithm fully suitable for practical everyday use. The industrial experts underlined the importance of continuous evaluation during applied research. Thus the incorporation of industrial best practices used for manual scheduling during the algorithm's design and full integration workers on the factory floor during the algorithm's introduction laid the foundation for a successful practical implementation.

Following the conclusion of research, all participating companies chose to purchase a fully developed scheduling software based on this algorithm and to use it as their primary means of performing Tactsequence Mapping and Production Scheduling.

# Chapter 7

# Conclusion

This marks the conclusion of the applied research process and thus of this thesis. As such this Chapter will both briefly summarize the results and provide an outlook on potential foci for future research.

In the introduction the situation tool manufacturers in Germany find themselves in was described as was the imperative to dramatically improve throughput-time and adherence to due-date. In light of many tool manufacturers' organizational structure, the need for a disruptive change in the tool manufacturing industry became evident. To confront the aforementioned challenges many tool manufacturers had been industrializing their processes and organizational structures. One highly promising aspect of industrialization was the introduction of a synchronized production system. Research, however, showed that those companies which had adopted this form of production had to deal with significant short term planning effort despite previous scientific work in this sector. This raised the first research research question regarding the necessary activities to effectively support the short-term planning of synchronized tool manufacturing.

As a next step this thesis' frame of reference was described by providing an in-depth overview of tools and tool manufacturing. After first examining industrialization and its effects on tool manufacturing the concept of synchronized production was explored in detail. Closer examination of prior work in this area revealed that no methods or algorithms existed to find suitable Tactsequences for jobs and existing scheduling algorithms were limited to processing only single Tactlines. Both of these gaps in the body of scientific knowledge caused significant manual effort for companies with synchronized production systems. This insight allowed for the deduction of two research questions, which targeted the development of algorithms to fill the identified gaps and allow for future automation of planning activities.

Subsequently, two algorithms referred to as "Tactsequence Mapping" and "Production

Scheduling" were presented. These can be implemented into software to almost completely automate short-term production planning activities. The proposed Tactsequence Mapping algorithms incorporate various approaches and algorithms from the areas of operations research and computer science to always yield optimal results. It was possible to translate the Production Scheduling problem into a combination of established problems from operations research. The complexity of several constituting sub-problems was, however, exceedingly high and precluded an analytical solution in the context of this thesis. Thus a heuristic was designed to provide solutions the scheduling problem by integrating both prior scientific work and input from industrial experts on production scheduling.

A combination of both a theoretical and practical evaluation confirmed the suitability of the developed algorithms for use in practice. In order to test the algorithms in an industrial setting, both were implemented as software prototypes. These prototypes were installed as the primary production planning tools in the tool manufacturing departments of three large German first-tier automotive suppliers. In this setting both algorithms exhibited acceptable running time on contemporary hardware and produced results, which were at least as good or better than those generated by the resident production planning experts. After completion of the research underlying this thesis, development of the software prototype was completed yielding a production planning system. This system is now installed permanently as the primary system used by all three companies, which contributed to the research project.

Thus this thesis has executed all of the activities necessary to successfully complete ULRICH's process of applied research. The presented results, however, open new perspectives for future research. One possibility of expanding the production planning algorithms is currently being explored in the *KoSyF* research project funded by the German Federal Ministry of Education and Research. The objective of this project is to develop a new form of collaborative production planning for synchronized production by opening the planning process to currently excluded stakeholders, e.g. workers on the factory floor. Another potential focus of future research is the development of a set of algorithms capable of adjusting a synchronized production system to a constantly changing product spectrum or machine complement. ZISKOVEN provided a very good guideline for the manual definition of Tactlines and the general setup of a synchronized production system. The continuous adaption of a production system is, however, deemed necessary by industrial experts and currently requires considerable manual effort. It is now possible to construct a reconfiguration algorithm for synchronized production systems, which can be run in a closed loop with the presented Tactsequence Mapping and Production Scheduling algorithms. This would allow said algorithms to evaluate reconfiguration alternatives regarding their impact on various performance metrics.

# List of Figures

# List of Tables

# Bibliography

Combinatorial Optimization. Berlin, Heidelberg: Springer Verlag.

Geschichte der Arbeit: Vom Alten Ägypten bis zur Gegenwart. Köln: Kiepenheuer & Witsch.

Intelligent Problem Solving. Methodologies and Approaches. Band 1821, Berlin, Heidelberg: Springer Verlag.

Operative Exzellenz im Werkzeug- und Formenbau. 1. Auflage. Aachen: Apprimus Verlag.

(2012): Mastering product complexity. Düsseldorf.

Handbuch Qualitätsmanagement. München: Hanser Verlag.

AARDAL, Karen/LENSTRA, Arjen K.: Hard Equality Constrained Integer Knapsacks. Springer Verlag, 350–366.

ABBOTT, Andrew/HRYCAK, Alexandra (1990): Measuring Resemblance in Sequence Data: An Optimal Matching Analysis of Musicians' Careers. American Journal of Sociology, 96, Nr. 1, 144–185.

ABELE, Eberhard/REINHART, Gunther (2011): Zukunft der Produktion: Herausforderungen, Forschungsfelder, Chancen. München: Hanser Verlag.

ABITEBOUL, S./HULL, Richard/VIANU, Victor (1995): Foundations of databases. Reading, Mass. and Wokingham: Addison-Wesley.

AUDI, Robert (1999): The Cambridge dictionary of philosophy. 2. Auflage. Cambridge: Cambridge University Press.

AUST, B. (1990): Ein Bewertungsverfahren für die Produktionsplanung bei auftragsorientierter Werkstattfertigung..

BALAS, Egon/ZEMEL, Eitan (1980): An Algorithm for Large Zero-One Knapsack Problems. Operations Research, 28, Nr. 5, 1130–1154.

BELLMAN, Richard (1957): Letter to the Editor—Comment on Dantzig's Paper on Discrete Variable Extremum Problems. Operations Research, 5, Nr. 5, 723–724.

BELLMAN, Richard (1972): Dynamic Programming..

BISPING, Marc (2003): Integrierte Produktstrukturen für den Werkzeugbau. Band 2003,5, Hannover: TEWISS Verlag.

BOLAND, Natashia et al. (2012): Clique-based facets for the precedence constrained knapsack problem. Mathematical Programming, 133, Nr. 1-2, 481–511.

BOOS, Wolfgang (2008): Methodik zur Gestaltung und Bewertung von modularen Werkzeugen. Band Band 2008,1, Aachen: Apprimus Verlag.

BOOS, Wolfgang et al. (2015): Tooling in China. Aachen.

BRAKEMEIER, Drik/JÄGER, Hans-Christian (2004): Schlanke Produktion: Als Wettbewerbsvorteil in globalen Märkten für Unternehmen in entwickelten Industrien. IM : Die Fachzeitschrift Für Information Management Und Consulting,, Nr. 19, 84–89.

BRANKAMP, Klaus: Organisation des Betriebsmittelbaus. Verlag Industrielle Organisation.

CAI, Xuan (2009): Canonical Coin Systems for CHANGE-MAKING Problems. In 2009 Ninth International Conference on Hybrid Intelligent Systems., 499–504.

CHANDRA, Ashok K./HIRSCHBERG, D. S./WONG, C. K. (1976): Approximate algorithms for some generalized knapsack problems. Theoretical Computer Science, 3, Nr. 3, 293–304.

CHANG, S. K./GILL, A. (1970): Algorithmic Solution of the Change-Making Problem. Journal of the ACM, 17, Nr. 1, 113–122.

CHEBIL, Khalil/KHEMAKHEM, Mahdi (2015): A dynamic programming algorithm for the Knapsack Problem with Setup. Computers & Operations Research, 64, 40–50.

CHEKURI, Chandra/KHANNA, Sanjeev (2005): A Polynomial Time Approximation Scheme for the Multiple Knapsack Problem. SIAM Journal on Computing, 35, Nr. 3, 713–728.

CLARKE, Constanze Anja (2002): Forms and functions of standardisation in production systems of the automotive industry: The case of Mercedes-Benz: Dissertation Freie Universität Berlin. Berlin.

CODD, Edgar F.: Further normalization of the data base relational model. Prentice Hall.

CODD, Edgar F.: Relational Completenes of Data Base Sublanguages. Prentice Hall.

CODD, Edgar F. (1970): A Relational Model of Data for Large Shared Data Banks. Communications of the ACM, 13, Nr. 6, 377–387.

CODD, Edgar F. (1971): Normalized data base structure. In Codd, Edgar F./Dean, Albert L. (Hrsg.): Proceedings of the 1971 ACM SIGFIDET Workshop on Data Description., 1–17.

CODD, Edgar F. (2009): Derivability, redundancy and consistency of relations stored in large data banks. SIGMOD Rec, 38, Nr. 1, 17–36.

COOPER, R. G. (1979): The Dimensions of Industrial New Product Success and Failure. Journal of Marketing, 43, Nr. 3, 93–103.

CORMEN, Thomas H. (2009): Introduction To Algorithms. 3. Auflage. Cambridge, Mass. and London: MIT Press.

CROCHEMORE, Maxime/HANCART, Christophe/LECROQ, Thierry (2007): Algorithms on Strings. Cambridge: Cambridge University Press.

CSENDES, Tibor (2001): New Subinterval Selection Criteria for Interval Global Optimization: Journal of Global Optimization. Journal of Global Optimization, 19, Nr. 3, 307–327.

DAMERAU, Frederick (1964): A Technique for Computer Detection and Correction of Spelling Errors. Communications of the ACM, 7, Nr. 3, 171–176.

DANTZIG, George B. (1957): Discrete-Variable Extremum Problems. Operations Research, 5, Nr. 2, 266–288.

DASGUPTA, Sanjoy/PAPADIMITRIOU, Christos H./VAZIRANI, Umesh Virkumar (2008): Algorithms. Boston: McGraw-Hill Higher Education.

DEVLIN, Keith J. (2003): Sets, functions, and logic. 3. Auflage. Boca Raton, Fla. and London: Chapman & Hall/CRC.

DI BENEDETTO, C. ANTHONY (1999): Identifying the Key Success Factors in New Product Launch. Journal of Product Innovation Management, 16, Nr. 6, 530–544.

DIUBIN, G. N./KORBUT, A. A. (2011): Greedy algorithms for the minimization knapsack problem: Average behavior. Journal of Computer and Systems Sciences International, 47, Nr. 1, 14–24.

DÖRING, Sebastian Thimo (2010): Konfliktmanagement in der technischen Auftragsabwicklung im Werkzeugbau. Band Bd. 2010,20, Aachen: Apprimus Verlag.

DÜR, Mirjam/STIX, Volker (2005): Probabilistic subproblem selection in branch-and-bound algorithms. Journal of Computational and Applied Mathematics, 182, Nr. 1, 67–80.

DYCKHOFF, Harald (1995): Grundzüge der Produktionswirtschaft. Berlin: Springer Verlag.

ERLEBACH, Thomas/KELLERER, Hans/PFERSCHY, Ulrich (2002): Approximating Multiobjective Knapsack Problems. Management Science, 48, Nr. 12, 1603–1612.

EUGÉNIA CAPTIVO, M. et al. (2003): Solving bicriteria 0–1 knapsack problems using a labeling algorithm. Computers & Operations Research, 30, Nr. 12, 1865–1886.

EVERSHEIM, Walter/KLOCKE, Fritz (1998): Werkzeugbau mit Zukunft: Strategie und Technologie. Berlin: Springer Verlag.

FAYARD, Didier/PLATEAU, Gérard (1994): An exact algorithm for the 0–1 collapsing knapsack problem. Discrete Applied Mathematics, 49, Nr. 1-3, 175–187.

FELDHUSEN, Jörg/GROTE, Karl-Heinrich (2013): Pahl/Beitz Konstruktionslehre: Methoden und Anwendung erfolgreicher Produktentwicklung. 8. Auflage. Berlin, Heidelberg: Springer Verlag.

FISCHER, Michael/WAGNER, Robert (1974): The String-to-String Correction Problem. Journal of the Association for Computing Machinery, 21, Nr. 1, 168–173.

FONG, Joseph (2015): Information systems reengineering, integration and normalization. 3. Auflage. Cham: Springer Verlag.

FORD, H./THESING, C. (1923): Mein Leben und Werk. 11. Auflage. Leipzig: Paul List Verlag.

FORD, H./THESING, C. (1930): Henry Ford und trotzdem Vorwärts! Leipzig: Paul List Verlag.

FRAENKEL, Abraham Adolf/BAR-HILLEL, Yehoshua/LÉVY, Azriel (1973): Foundations of set theory. Band 67, 2. Auflage. Amsterdam: Elsevier Science B.V..

FRÉVILLE, Arnaud (2004): The multidimensional 0–1 knapsack problem: An overview. European Journal of Operational Research, 155, Nr. 1, 1–21.

FRICK, Lutz (2006): Erfolgreiche Geschäftsmodelle im Werkzeugbau. Band 2006,5, Aachen: Shaker Verlag.

FRICKER, Ingo Christian (2005): Strategische Stringenz im Werkzeug- und Formenbau. Band Bd. 2005,16, Aachen: Shaker Verlag.

GALLO, G./HAMMER, P. L./SIMEONE, B.: Quadratic knapsack problems. Springer Verlag, 132–149.

GAREY, M. R./JOHNSON, David S. (1978): Strong NP-Completeness Results: Motivation, Examples and Implications. Journal of the ACM, 25, Nr. 3, 499–508.

GAREY, Michael R./JOHNSON, David S. (1979): Computers and intractability: A guide to the theory of NP-completeness / Michael R. Garey, David S. Johnson. San Francisco, California: Freeman.

GENS, GEORGII V. AND EUGENII V. LEVNER (1979): Computational complexity of approximation algorithms for combinatorial problems. Mathematical Foundations of Computer Science,, 292–300.

GIEHLER, Florian (2010): Erhöhung der Planungsproduktivität am Beispiel der Auftragsabwicklung im Werkzeugbau. Band [2010,16], 1. Auflage. Aachen: Apprimus Verlag.

GOLDREICH, Oded (2010): P, NP, and NP-completeness: The basics of computational complexity. Cambridge and New York: Cambridge University Press.

GÜNTZER, Michael M./JUNGNICKEL, Dieter (2000): Approximate minimization algorithms for the 0/1 Knapsack and Subset-Sum Problem. Operations Research Letters, 26, Nr. 2, 55–66.

GUSFIELD, Dan (2009): Algorithms on Strings, Trees, and Sequences: Computer Science and Computational Biology. 12. Auflage. Cambridge: Cambridge University Press.

GUTENBERG, Erich (1960): Grundlagen der Betriebswirtschaftslehre: Band 1: Die Produktion. 5. Auflage. Berlin and Heidelberg: Springer Verlag.

HAMMING, Richard Wesley (1950): Error Detecting and Error Correcting Codes. Bell System Technical Journal, 29, Nr. 2, 147–160.

HARTMANIS, Juris (1989): Godel, von Neumann and the P=? NP Problem. Bulletin of the European Association for Theoretical Computer Science,, Nr. 38, 101–107.

HEMPEL, Carl G./OPPENHEIM, Paul (1948): Studies in the logic of explanation. Philosophy of Science, 15, Nr. 2, 135–175.

HERLITZ, Jörg (1995): Lean Management als Wettbewerbsstrategie im deutschen Werkzeugmaschinenbau. Berlin: Freie Universität Berlin.

HILL, W./ULRICH, Hans: Wissenschaftliche Aspekte ausgewählter betriebswirtschaftlicher Konzeptionen. Vahlen.

HINSEL, Christian: Synchrone Fließfertigung: Auch in einem Werkzeugbau der Massivumformungen? Fraunhofer-Institut für Produktionstechnologie, IPT.

HOLMES, J./RUTHERFORD, T./FITZGIBBON, S.: Innovation in the Automotive Tool, Die and Mold Industry: A Case Study of the Windsor-Essex Region. McGill-Queens University Press.

HOYER, H./KNUTH, M. (1976): Die teilautonome Gruppe. Strategie des Kapitals oder Chance für die Arbeiter? In Kursbuch Nr. 43. Berlin: Kursbuch/Rotbuch-Verlag.

HROMKOVIČ, Juraj (2004): Algorithmics for Hard Problems: Introduction to Combinatorial Optimization, Randomization, Approximation, and Heuristics. Second edition Auflage. Berlin, Heidelberg: Springer Verlag.

HUNG, Ming S./FISK, John C. (1978): An algorithm for 0-1 multiple-knapsack problems. Naval Research Logistics Quarterly, 25, Nr. 3, 571–579.

HÜTTMEIR, Andreas et al. (2009): Trading off between heijunka and just-in-sequence. International Journal of Production Economics, 118, Nr. 2, 501–507.

IBARRA, Oscar H./KIM, Chul E. (1975): Fast Approximation Algorithms for the Knapsack and Sum of Subset Problems. J. ACM, 22, Nr. 4, 463–468.

IIDA, Hiroshi (1999): A Note on the Max-Min 0-1 Knapsack Problem. Journal of Combinatorial Optimization, 3, Nr. 1, 89–94.

JANSEN, Klaus/MARGRAF, Marian (2008): Approximative Algorithmen und Nichtapproximierbarkeit. Berlin and New York, NY: De Gruyter.

JAYARĀMAN, Pi. (2007): Management Icons. 1. Auflage. New Delhi: Excel Books.

JOHNSON, David S./NIEMI, K. A. (1983): On Knapsacks, Partitions, and a New Dynamic Programming Technique for Trees. Mathematics of Operations Research, 8, Nr. 1, 1–14.

JUNGNICKEL, D. (2005): Graphs, networks and algorithms. Band 5, 2. Auflage. Berlin and London: Springer Verlag.

KARP, Richard M.: Reducibility among Combinatorial Problems. Springer Verlag, 85–103.

KEDAR, S. (2009): Database Management System. Technical Publications.

KELLERER, Hans/PFERSCHY, Ulrich/PISINGER, David (2004): Knapsack Problems. Berlin, Heidelberg: Springer Verlag.

KESSLER, Heinrich/WINKELHOFER, Georg (2004): Projektmanagement: Leitfaden zur Steuerung und Führung von Projekten ; mit 42 Tabellen. 4. Auflage. Heidelberg: Springer Verlag.

KLOCKE, Fritz/SCHUH, Günther (2005): Zukunftsstudie Werkzeug- und Formenbau. Aachen: Werkzeugmaschinenlabor WZL der RWTH Aachen.

KLOTZBACH, Christoph (2007): Gestaltungsmodell für den industriellen Werkzeugbau. Band 2007,1, Aachen: Shaker Verlag.

KNUTH, Donald E./MORRIS, JR., JAMES H./PRATT, Vaughan R. (1977): Fast Pattern Matching in Strings. SIAM Journal on Computing, 6, Nr. 2, 323–350.

KOEPPEN, Mario/SCHAEFER, Gerald/ABRAHAM, Ajith (2011): Intelligent Computational Optimization in Engineering: Techniques & Applications. Springer Verlag.

KOLESAR, Peter (1967): A Branch and Bound Algorithm for the Knapsack Problem. Management Science, 13, Nr. 9, 723–735.

KOLLIOPOULOS, Stavros G./STEINER, George (2007): Partially ordered knapsack and applications to scheduling. Discrete Applied Mathematics, 155, Nr. 8, 889–897.

KORTE, Bernhard/VYGEN, Jens (2008): Combinatorial optimization: Theory and algorithms. Band 21, 3. Auflage. Berlin [u.a.]: Springer Verlag.

KOZIELSKI, Stefan (2010): Integratives Kennzahlensystem für den Werkzeugbau. Band 2010,19, Aachen: Apprimus Verlag.

KRAFCIK, J., F. (1988): Triumph of the lean production system..

KRESS, Moshe/PENN, Michal/POLUKAROV, Maria (2007): The minmax multidimensional knapsack problem with application to a chance-constrained problem. Naval Research Logistics, 54, Nr. 6, 656–666.

KROMREY, Helmut (1991): Empirische Sozialforschung. 5. Auflage. Opladen: Leske Verlag.

KUBICEK, Herbert (1976): Heuristische Bezugsrahmen und heuristisch angelegte Forschungsdesign als Elemente einer Konstruktionsstrategie empirischer Forschung. Band Nr. 16, Berlin: Institut für Unternehmungsführung im Fachbereich Wirtschaftswissenschaften der freien Universität Berlin.

KUHN, Thomas S. (2001): Die Struktur wissenschaftlicher Revolutionen. Band 25, 15. Auflage. Frankfurt am Main: Suhrkamp.

KULIK, Ariel/SHACHNAI, Hadas (2010): There is no EPTAS for two-dimensional knapsack. Information Processing Letters, 110, Nr. 16, 707–710.

KYNN, Mary (2008): The 'heuristics and biases' bias in expert elicitation. Journal of the Royal Statistical Society: Series A, 171, Nr. 1, 239–264.

LAND, A. H./DOIG, A. G. (1960): An Automatic Method of Solving Discrete Programming Problems. Econometrica, 28, Nr. 3, 497.

LAWLER, Eugene L. (1979): Fast Approximation Algorithms for Knapsack Problems. Mathematics of Operations Research, 4, Nr. 4, 339–356.

LEVENSHTEIN, Vladimir (1966): Binary Codes Capable of Correcting Deletions, Insertions, and Reversals. Soviet Physics Doklady 10, Nr. 8.

LI, Yuefeng/LOOI, Mark/ZHONG, Ning (2006): Advances in intelligent IT: Active media technology 2006. Band v. 138, Amsterdam: IOS Press.

LÖCHEL, H. (2013): Mikroökonomik: Haushalte, Unternehmen, Märkte. Gabler Verlag.

LUEKER, G. S. (1975): Two NP-complete problems in nonnegative integer programming. Band 178, Technical Report. Computer Science Laboratory, Princeton University.

LUUS, Rein (2000): Iterative dynamic programming. Band 110, Boca Raton: Chapman & Hall/CRC.

MAJSTER, Mila E./REISER, Angelika (1980): Efficient On-Line Construction and Correction of Position Trees. SIAM Journal on Computing, 9, Nr. 4, 785–807.

MALIK, Fredmund (1992): Strategie des Managements komplexer Systeme: Ein Beitrag zur Management-Kybernetik evolutionärer Systeme. 4. Auflage. Bern and Stuttgart and Wien: Haupt.

MARCZINSKI, G. (2008): Lean Production 2.0: Lean Production 2.0 - Statt entweder ERP oder Lean Manufacturing gilt sowohl als auch! Zeitschrift für den Betriebswissenschaftlichen Fabrikbetrieb,, Nr. 11, 804–808.

MARTELLO, Silvano/TOTH, Paolo (1977): An upper bound for the zero-one knapsack problem and a branch and bound algorithm. European Journal of Operational Research, 1, Nr. 3, 169–175.

MARTELLO, Silvano/TOTH, Paolo (1980): Optimal and canonical solutions of the change making problem. European Journal of Operational Research, 4, Nr. 5, 322–329.

MARTELLO, Silvano/TOTH, Paolo (1990): Knapsack problems: Algorithms and computer implementations..

MATHEWS, G. B. (1897): On the partition of numbers. Proceedings of the London Mathematical Society,, Nr. 28, 486–490.

MCLAY, Laura A./JACOBSON, Sheldon H. (2007): Integer knapsack problems with set-up weights. Computational Optimization and Applications, 37, Nr. 1, 35–47.

MCMILLAN, Michael (2007): Data Structures and Algorithms Using C#. Cambridge University Press.

MENEZES, J (2004): European Mold Making: Towards a New Competitive Positioning. In Schuh, Günther (Edit.): Proceedings of the 4th International Colloquium Tool and Die Making for the Future. Aachen: Apprimus Verlag.

MIETTINEN, Kaisa (2012): Nonlinear Multiobjective Optimization. Springer Verlag.

MOKYR, Joel (1999): The British industrial revolution: An economic perspective / edited by Joel Mokyr. 2. Auflage. Boulder, Colo. and Oxford: Westview Press.

MORENO, Eduardo/ESPINOZA, Daniel/GOYCOOLEA, Marcos (2010): Large-scale multi-period precedence constrained knapsack problem: A mining application. Electronic Notes in Discrete Mathematics, 36, 407–414.

MOSER, Klaus (2007): Mass customization strategies: Development of a competence-based framework for identifying different mass customization strategies. [Place of publication not identified]: Klaus Moser.

NEAPOLITAN, Richard E. (2015): Foundations of algorithms. Fifth edition Auflage. Burlington MA: Jones & Bartlett Learning.

NEDESS, Christian/HAUER, Christian (1997): Organisation des Produktionsprozesses. Stuttgart: Teubner.

PADAWITZ, Peter (1992): Deduction and declarative programming. Band 28, Cambridge [England] and New York: Cambridge University Press.

PANDEY, H. M. (2008): Design Analysis and Algorithm. 1. Auflage. University Science Press.

PARDALOS, P. M./MIGDALAS, A./PITSOULIS, L. (2008): Pareto Optimality, Game Theory and Equilibria. Springer Verlag.

PARKER, R. Gary/RARDIN, Ronald L. (1988): Discrete Optimization..

PAULINYI, Á: Die Entstehung des Fabriksystems in Großbrittannien. Kiepenheuer & Witsch.

PEARSON, David (1994): A polynomial-time algorithm for the change-making problem. Operations Research Letters, 33, Nr. 3, 231–234.

PFERSCHY, Ulrich/PISINGER, David/WOEGINGER, Gerhard J. (1997): Simple but efficient approaches for the collapsing knapsack problem. Discrete Applied Mathematics, 77, Nr. 3, 271–280.

PINTO, Telmo et al. (2015): Solving the Multiscenario Max-Min Knapsack Problem Exactly with Column Generation and Branch-and-Bound. Mathematical Problems in Engineering, 2015, 1–11.

PISINGER, David (1995): A minimal algorithm for the multiple-choice knapsack problem. European Journal of Operational Research, 83, Nr. 2, 394–410.

PISINGER, David (2007): The quadratic knapsack problem—a survey. Discrete Applied Mathematics, 155, Nr. 5, 623–648.

PITSCH, Martin et al. (2015): Getaktete Fertigung im Werkzeugbau. 1. Auflage. Aachen: Werkzeugmaschinenlabor WZL der RWTH Aachen.

PONNIAH, Paulraj (2007): Data modeling fundamentals: A practical guide for IT professionals / Paulraj Ponniah. Hoboken, N.J..

POOLE, D. L./MACKWORTH, A. K. (2010): Artificial Intelligence: Foundations of Computational Agents. Cambridge University Press.

POSNER, Marc E./GUIGNARD, Monique (1978): The Collapsing 0–1 Knapsack Problem. Mathematical Programming, 15, Nr. 1, 155–161.

PRINZIE, Anita/VAN DEN POEL, Dirk (2006): Incorporating Sequential Information into Traditional Classification Models by Using an Element/Position-sensitive SAM. Decision Support Systems, 42, Nr. 2, 508–526.

PUCHINGER, Jakob/RAIDL, Günther R./PFERSCHY, Ulrich (2010): The Multidimensional Knapsack Problem: Structure and Algorithms. INFORMS Journal on Computing, 22, Nr. 2, 250–265.

RADER JR., David J./WOEGINGER, Gerhard J. (2002): The quadratic 0–1 knapsack problem with series–parallel support. Operations Research Letters, 30, Nr. 3, 159–166.

RAM, Balasubramanian/SARIN, Sanjiv (1988): An Algorithm for the 0-1 Equality Knapsack Problem. The Journal of the Operational Research Society, 39, Nr. 11, 1045.

REVELLE, Jack B. (2001): Manufacturing handbook of best practices: An innovation, productivity, and quality focus / edited by Jack B. ReVelle. Boca Raton, FL: St. Lucie Press.

RISSE, Jörg (2003): Time-to-Market-Management in der Automobilindustrie: Ein Gestaltungsrahmen für ein logistikorientiertes Anlaufmanagement. Band Bd. 4, 1. Auflage. Bern: Haupt.

RITZ, Peter (1999): Bewertung technischer Änderungen im Werkzeugbau. Band Bd. 99,28, Aachen: Shaker Verlag.

ROTHLAUF, Franz (2011): Design of modern heuristics: Principles and application. Heidelberg and New York: Springer Verlag.

SALOMON, D./MOTTA, Giovanni (2010): Handbook of data compression. 5. Auflage. London: Springer Verlag.

SCHANZ, G.: Wissenschaftstheoretische Grundfragen der Führungsforschung. Poeschel, 2189–2197.

SCHLEYER, Carsten (2006): Erfolgreiches Kooperationsmanagement im Werkzeugbau. Band Bd. 2006,26, Aachen: Shaker Verlag.

SCHLÜTER, Helmut (23.8.2006): Vorkalkulationsmethoden von Spritzgießwerkzeugen: Auf Basis der Ähnlichkeitsbetrachtung. Würzburg.

SCHRIJVER, Alexander (2003): Combinatorial Optimization: Polyhedra and Efficiency. Band 24, Berlin and London: Springer Verlag.

SCHRÖDER, Carsten (2003): Aufbau hierarchiearmer Produktionsnetzwerke: Technologiestrategische Option und organisatorische Gestaltungsaufgabe. Stuttgart: Fraunhofer IRB Verlag.

SCHUH, Günther (2012): Handbuch Produktion und Management 3. 2. Auflage. Berlin: Springer Verlag.

SCHUH, Günther/KLOTZBACH, Christoph (29.11.2005): Situation of the Moldmaking Industry and Success Factors for its Development. Wiesbaden.

SCHUURMAN, P./WOEGINGER, R. H. (2007): Approximation Schemes. In Möhring, R. H. et al. (Hrsg.): Lectures on Scheduling. yet to appear.

SHACHNAI, Hadas/TAMIR, Tami: Polynomial-Time Approximation Schemes. Chapman & Hall/CRC, 9–1–9–21.

SHARMA, Kal Renganathan (2009): Bioinformatics: Sequence Alignment and Markov Models. New York: McGraw-Hill Professional.

SHAW, Dong X./CHO, Geon (1998): The critical-item, upper bounds, and a branch-and-bound algorithm for the tree knapsack problem. Networks, 31, Nr. 4, 205–216.

SHAW, Dong X./CHO, Geon/CHANG, Hsuliang (1997): A depth–first dynamic programming procedure for the extended tree knapsack problem in local access network design. Telecommunication Systems, 7, Nr. 1, 29–43.

SKIENA, Steven S. (2008): The algorithm design manual. 2. Auflage. London: Springer Verlag.

SNYMAN, Jan A. (2005): Practical mathematical optimization: An introduction to basic optimization theory and classical and new gradient-based algorithms / by Jan A. Snyman. Band v. 97, New York: Springer Verlag.

SPATH, Dieter et al. (2001): Vom Markt zum Markt: Produktentstehung als zyklischer Prozess. Stuttgart: Logis.

SPENNEMANN, Frank (2001): Gestaltung von Organisationsstrukturen im Werkzeugbau. Band Bd. 2001,14, Aachen: Shaker Verlag.

STATNIKOV, R. B./MATUSOV, Joseph B. (1995): Multicriteria optimization and engineering. New York: Chapman & Hall/CRC.

STEPHEN A. COOK (1971): The complexity of theorem-proving procedures. In Proceedings of the third annual ACM symposium on Theory of computing. Shaker Heights, Ohio, USA: ACM, 151–158.

SUMATHI, S./ESAKKIRAJAN, S. (2007): Fundamentals of relational database management systems. Band v. 47, Berlin and London: Springer Verlag.

SUNG, W. K. (2009): Algorithms in Bioinformatics: A Practical Introduction. CRC Press.

TANIGUCHI, Fumiaki/YAMADA, Takeo/KATAOKA, Seiji (2008): Heuristic and exact algorithms for the max–min optimization of the multi-scenario knapsack problem. Computers & Operations Research, 35, Nr. 6, 2034–2048.

TAYLOR, Frederick Winslow (1919): Die Grundsätze wissenschaftlicher Betriebsführung. München: Oldenbourg.

TICHY, Walter (1984): The String-to-String Correction Problem with Block Moves. ACM Transactions on Computer Systems, 2, Nr. 4, 309–321.

TOMCZAK, T. (1992): Forschungsmethoden in der Marketingwissenschaft - Ein Plädoyer für den qualitativen Forschungsansatz. Marketing ZFP, 2, Nr. 2, 129.

ULRICH, Hans (1982): Anwendungsorientierte Wissenschaft. Die Unternehmung, 36, Nr. 1, 1–10.

VAN DYKE PARUNAK, H. (1991): Characterizing the Manufacturing Scheduling Problem. Journal of Manufacturing Systems, 10, Nr. 3, 241–259.

VAN LEEUWEN, J. (1994): Handbook of Theoretical Computer Science: Vol. A: Algorithms and Complexity. 1. Auflage. Cambridge, Massachusetts: MIT Press.

VAZIRANI, Vijay V. (2001): Approximation algorithms. Berlin and London: Springer Verlag.

VDI (1978): Begriffe für die Produktionsplanung und -steuerung. Düsseldorf.

**VDMA (2014):** Statistisches Handbuch für den Machinenbau: Ausgabe 2014. VDMA Verlag.

**VOLGENANT, A./MARSMAN, S. (1998):** A Core Approach to the 0-1 Equality Knapsack Problem. The Journal of the Operational Research Society, 49, Nr. 1, 86.

**WAGNER, Peter (2004):** Kundenorientierung: Der Königsweg zum Unternehmenserfolg. Band 28, 3. Auflage. Renningen: Expert Verlag.

**WANG, John X. (2011):** Lean manufacturing: Business bottom-line based. Boca Raton, FL: CRC Press.

**WEINBERG, Franz (1968):** Einführung in die Methode Branch and Bound: Unterlagen für einen Kurs des Instituts für Operations Research der ETH, Zürich. Band 4, Berlin and Heidelberg and New York: Springer Verlag.

**WESTEKEMPER, Markus (2002):** Methodik zur Angebotspreisbildung am Beispiel des Werkzeug- und Formenbaus. Aachen: Shaker Verlag.

**WESTFECHTEL, Bernhard (1999):** Models and Tools for Managing Development Processes. Band 1646, Berlin: Springer Verlag.

**WIENDAHL, Hans-Peter/KUPRAT, Thomas/AHRENS, Volker:** Logistikgerechte Gestaltung von Produktionsstrukturen auf der Basis von Betriebskennlinien — Theorie und praktische Anwendung in der Metall- und Elektroindustrie. Gabler Verlag, 221–248.

**WILLRETT, H. (2011):** Im Takt den Vorsprung sichern. Industrieanzeiger 133, Nr. 5.

**WOEPPEL, Mark J. (2000):** The manufacturer's guide to implementing the theory of constraints. Boca Raton, FL and London: St. Lucie Press.

**WOMACK, James P./JONES, Daniel T./ROOS, Daniel (1990):** The machine that changed the world: The story of lean production–Toyota's secret weapon in the global car wars that is revolutionizing world industry. New York: Rawson.

**WOMACK, James P./JONES, Daniel T./ROOS, Daniel (1991):** Die zweite Revolution in der Autoindustrie: Konsequenzen aus der weltweiten Studie aus dem Massachusetts Institute of Technology. 2. Auflage. Frankfurt am Main: Campus Verlag.

**WU, J./SRIKANTHAN, T. (2006):** An efficient algorithm for the collapsing knapsack problem. Information Sciences, 176, Nr. 12, 1739–1751.

YANG, Xin-She (2010): Engineering optimization: An Introduction with Metaheuristic Applications. Hoboken, N.J.: John Wiley.

YOSHIMURA, Masataka (2010): System design optimization for product manufacturing. London: Springer Verlag.

YOU, Byungjun/YAMADA, Takeo (2007): A pegging approach to the precedence-constrained knapsack problem. European Journal of Operational Research, 183, Nr. 2, 618–632.

YU, Gang (1996): On the Max-Min 0-1 Knapsack Problem with Robust Optimization Applications. Operations Research, 44, Nr. 2, 407–415.

ZHANG, W. (2012): State-Space Search: Algorithms, Complexity, Extensions, and Applications. Springer Verlag.

ZISKOVEN, Hagen (2013): Methodik zur Gestaltung und Auftragseinplanung einer getakteten Fertigung im Werkzeugbau. Band Band 2013,17, Aachen: Apprimus Verlag.

ZWANZIG, Florian (2010): Taktung der Unikatfertigung am Beispiel des Werkzeugbaus. Band Band 2010,1, Aachen: Apprimus Verlag.

ŌNO, Taiichi (1993): Das Toyota-Produktionssystem. Frankfurt/Main and New York: Campus Verlag.

# Appendix A

# Supervised Theses

During the research underlying this dissertation the following project, bachelor, master and diploma theses were supervised:

Matthias Scholer: Concept design of a methodology for the process data acquisition and quality level assessment of PCB suppliers, Bachelor Thesis RWTH Aachen, 2012

Lukas Vogel: Development of a concept to ensure transparency within Engineering Change Management, Bachelor Thesis RWTH Aachen, 2012

Bartosch Haratyk: Konzeption eines Modells zur Verwendung mittels Embedded Toolkits gewonnener Informationen aus der Produktnutzungsphase entlang des Produktlebenszyklus, Diploma Thesis RWTH Aachen, 2012

Matthis Laass: Entwicklung eines Software-Frameworks zur Identifikation und Bewertung von Nachrichten in sozialen Medien, Master Thesis FH Aachen, 2012

Frederik Dietz, Johannes Graf, Dennis Grunert: Konzeption, Aufbau und Implementierung eines Prototyps für den domänenübergreifenden Einsatz von Embedded Toolkits in den Bereichen Smart-Phone und Smart-Car, Project Thesis RWTH Aachen, 2013

Laura Agnes Pilsel: Entwicklung eines Algorithmus zur Auslegung von Linien einer getakteten Fertigung im Werkzeugbau, Master Thesis RWTH Aachen, 2014.